Causality Rules (Second Edition)

Dispersion theory in non-elementary particle physics

Online at: https://doi.org/10.1088/978-0-7503-3431-0

Causality Rules (Second Edition)

Dispersion theory in non-elementary particle physics

Vladimir Pascalutsa

Department of Physics, Mathematics and Computer Science, Johannes Gutenberg University of Mainz, Staudingerweg 7, 55128 Mainz, Germany

IOP Publishing, Bristol, UK

ISBN 978-0-7503-3431-0 (ebook)
ISBN 978-0-7503-3429-7 (print)
ISBN 978-0-7503-3432-7 (myPrint)
ISBN 978-0-7503-3430-3 (mobi)

DOI 10.1088/978-0-7503-3431-0

Version: 20240401

IOP ebooks

British Library Cataloguing-in-Publication Data: A catalogue record for this book is available from the British Library.

Published by IOP Publishing, wholly owned by The Institute of Physics, London

IOP Publishing, No.2 The Distillery, Glassfields, Avon Street, Bristol, BS2 0GR, UK

US Office: IOP Publishing, Inc., 190 North Independence Mall West, Suite 601, Philadelphia, PA 19106, USA

Contents

Preface

It is the theory that decides what we can observe.

–Albert Einstein

This book is about powerful relations due to causality, often in combination with other general principles, such as unitarity and space–time symmetries. These general relations are widely used in many fields of physics, from optics and atomic theory to gaining insights into quantum gravity. Yet, they are rarely a part of the standard curriculum; perhaps with the exception of the Kramers–Kronig relation which is the first known relation of this kind.

There are many more, uncovered here in the context of timely applications in particle, nuclear and atomic physics. The presentation for the first edition was rather brief, appropriately for the 'IOP Concise Physics' series. This is a significantly updated, corrected and expanded second edition. The level will suit master's and graduate students who, in addition to general physics courses, are familiar with complex analysis and elements of quantum field theory. The first two or three chapters should even be accessible at the undergraduate level. The last chapters are, in contrast, more technical and advanced. I can only hope that my students will be motivated enough to read through all of it.

Acknowledgements

Manuscripts don't burn

—Mikhail Bulgakov *Master and Margarita*

It is a pleasure to acknowledge the people who influenced my work on this book. First and foremost, Marc Vanderhaeghen—my colleague and dear friend, with whom I worked on this subject for more than 15 years. Further in the past, I was greatly influenced by my PhD advisor at ITF Utrecht, Professor John Tjon. His PhD advisor, Professor Nico van Kampen, and his PhD advisor, Professor Hendrik Kramers, were at the origins of this subject. Speaking of PhD students, I am greatly indebted to one of my own, Dr Franziska Hagelstein, who has made many useful corrections to the original manuscript. Another former student, Dr Oleksii Gryniuk, has kindly helped me with some of the figures.

Last but not least, I would like to thank Marc Pfirter for his warm and generous hospitality at the magnificent Villa Ferrette in St Blasien, where most of this book was written in an inspiring and enchanting setting. Regrettably, the Villa has since been burned to the ground in a devastating accident. This book is offered as a humble tribute to the lost masterpiece of Júgendstil.

Author biography

Vladimir Pascalutsa

Dr Vladimir Pascalutsa is a senior scientist at Johannes Gutenberg University of Mainz. He is an accomplished theoretical physicist, having more than 100 publications in the field of subatomic physics. He teaches advanced courses in quantum field theory and particle physics; supervises the research of MSc and PhD students.

He obtained his PhD in 1998 from the Institute for Theoretical Physics at Utrecht University, under the supervision of Professor John Tjon. This PhD lineage extends back to van Kampen, Kramers, Ehrenfest, Boltzmann, Stefan.

Subsequently, he held a series of research positions at universities and laboratories across the globe: NIKHEF (Amsterdam), Flinders University (Adelaide), Ohio University, the College of William and Mary, Jefferson Lab, the ECT* Trento, and since 2008, the JGU Mainz.

Chapter 1

Introduction

Time is what keeps everything from happening at once. —*Ray Cummings*
Time is an illusion. —*Albert Einstein*

Causality is a property of time. The effect comes *after* the cause, and vice versa. Therefore, the present can only be affected by the past, and only affects the future. This is not as simple in relativity, where time and space are intertwined. The concept of *before* and *after* becomes observer-dependent, and even though causality is maintained within each observer's reference frame, they may disagree with one another on what is the cause and the effect. It is rather mind-boggling to imagine such situations, but clearly the notion of causality which relies solely on time is ill-defined in relativity, special or general alike. As the cause–effect relationship is blurred, causality becomes only a statement that everything happens within the light cone, which is equivalent to the principal postulate of relativity that nothing travels faster than light.

This can be extended to relativistic quantum theory. Once the quantum mechanics and special relativity merge into quantum field theory (QFT), the quantum fluctuations are confined to the light cone. The latter condition is referred to as *microcausality*.

To date, there is no fully consistent merger of quantum mechanics and general relativity, viz., quantum gravity. Therefore, there is room for speculation on how causality is embedded therein. Nonetheless, a rather productive approach to quantum gravity is to deduce its properties from microcausality. This is accomplished by utilizing dispersion relations (S-matrix analyticity), which are the direct outcomes of causality.

This method of gaining insights into an unknown theory by using dispersion relations is not new. The subject is about a century old. Its beginnings are marked by the celebrated works of de Kronig [1] and Kramers [2] in the mid-1920s. The field of particle physics boomed in the 1950s and 60s, with the development of QFT and rapid advancement of the physics of strong interactions. Amid these developments, Gell-Mann, Goldberger and Thirring [3] derived relativistic Kramers–Kronig type relations for forward Compton scattering, which paved the way to the profound relations called *sum rules*. The prominent examples to be discussed further in this book include the sum rules of Baldin [4], Gerasimov–Drell–Hearn (GDH) [5, 6], Burkhardt–Cottingham (BC) [7], and Schwinger [8].

For a while, dispersion theory was the main approach for the physics of the strong interaction, until, in the 1970s, quantum chromodynamics (QCD) was established as the fundamental quantum theory of the strong interaction. However, the non-perturbative nature of QCD at low energies still makes the *ab initio* calculations extremely difficult and some progress has only been achieved in large computer simulations based on lattice QCD. The dispersive methods remain to be used, complementing the QCD calculations.

More specifically, the dispersion relations and sum rules are instrumental in many of the data-driven evaluations of hadronic and nuclear corrections. As dispersion relations provide general, model-independent relations amongst different observable quantities, they allow for an assessment of one quantity using a measurement of another.

A famous example is the $g-2$ of the electron and muon (i.e. the anomalous magnetic moment), considered in chapter 9. There the leading hadronic effects are assessed via a dispersion relation using the experimental cross sections of $e^+e^-\to$ hadrons as input [9–12]. Currently, there is, in fact, a significant disagreement between the data-driven and lattice-QCD evaluations of this effect.

Another important example comes from the nuclear-structure effects in atoms. These effects are especially interesting in light muonic atoms and ions, i.e. hydrogen-like systems where an electron is replaced by a muon. The use of dispersion relations in these atomic calculations will be considered in chapter 10.

Prior to coming to these practical applications, there are chapters which introduce the basic concepts and physical interpretations. To define the scope of this presentation more precisely, let us consider a generic dispersion relation:

$$\mathrm{Re} f(\omega) = \frac{1}{\pi} \fint_{-\infty}^{\infty} d\omega' \, \frac{\mathrm{Im} f(\omega')}{\omega' - \omega}. \tag{1.1}$$

It involves a complex amplitude, $f = \mathrm{Re} f + i\, \mathrm{Im} f$, describing usually the scattering and absorption of light or, generally, any signal or wave, with frequency ω; the slashed integral stands for the principal-value integration. Depending on the field of application, the amplitude f can be different things: the refraction index (chapter 3), the amplitude(s) of Compton scattering (chapter 4), doubly virtual Compton scattering (chapters 5, 8, and 10), light-by-light scattering (chapters 6 and 8), or vacuum polarization (chapter 9). In chapter 2, it is kept as general as possible, with the focus on the basic principles.

References

[1] Kronig R D 1926 On the theory of dispersion of x-rays *J. Opt. Soc. Am.* **12** 547–56

[2] Kramers H A 1927 *Atti Congr. Int. Fis. Como* **2** 545

[3] Gell-Mann M, Goldberger M and Thirring W E 1954 Use of causality conditions in quantum theory *Phys. Rev.* **95** 1612–27

[4] Baldin A M 1960 Polarizability of nucleons *Nucl. Phys.* **18** 310–7

[5] Gerasimov S 1966 A sum rule for magnetic moments and the damping of the nucleon magnetic moment in nuclei *Sov. J. Nucl. Phys.* **2** 430–3

[6] Drell S and Hearn A C 1966 Exact sum rule for nucleon magnetic moments *Phys. Rev. Lett.* **16** 908–11

[7] Burkhardt H and Cottingham W N 1970 Sum rules for forward virtual Compton scattering *Annals Phys.* **56** 453–63

[8] Schwinger J 1975 Source theory discussion of deep inelastic scattering with polarized particles *Proc. Natl Acad. Sci. USA* **72** 1559–63

[9] Melnikov K and Vainshtein A 2007 *Theory of the Muon Anomalous Magnetic Moment* (Springer Tracts in Modern Physics) (Berlin: Springer)

[10] Jegerlehner F and Nyffeler A 2009 The muon $g-2$ *Phys. Rep.* **477** 1–110

[11] Jegerlehner F 2017 *The Anomalous Magnetic Moment of the Muon* (Springer Tracts in Modern Physics vol 274) (Berlin: Springer)

[12] Aoyama T *et al* 2020 The anomalous magnetic moment of the muon in the standard model *Phys. Rep.* **887** 1–166

Chapter 2

Some rules for sum rules

In this chapter we describe the connection between causality and analyticity which underpins all dispersion relations and show the math that goes into their derivation. We cover the arguments of crossing symmetry, unitarity and low-energy theorems on the path toward deriving the sum rules. We close with a discussion of the convergence issues: subtractions, renormalization and regularization, followed by an approximate sum rule for the proton charge.

2.1 Causality and analyticity

Although not explicitly mentioned in the original papers by Kramers and Kronig, the physical principle behind a dispersion relation, such as (1.1), is *causality*. The first published report of the relation to causality can be traced back to a proceedings contribution of Kronig [1]. A more rigorous derivation and comprehensive review of causality constraints was presented by van Kampen [2, 3] (a PhD student of Kramers at the time).

In a causal world, any signal can be a superposition of signals from the past but not the future. That is, a signal at time t is influenced by signals at times $t' \leqslant t$. In the terminology of wave mechanics and field theory, the Green's functions, which describe the signal (or wave) propagation, absorption and scattering, must vanish for $t' > t$. This in turn infers the *analyticity* of the Fourier transform of the Green's functions. The Fourier-transform conjugate variable to t is the frequency ω, and hence, causality constrains the analytic properties of the ω dependence.

In a relativistic theory, the causality condition is somewhat more involved, for obvious reasons: time is relative. In this case one requires that signals propagate within the *light cone*, and hence, the Green's functions vanish for space-like separations. This means that the propagation speed is limited by the speed of light

doi:10.1088/978-0-7503-3431-0ch2

in a vacuum[1]. This principle applied to quantum fluctuations is referred to as *micro-causality*.

It is important to recognize that the maximal speed limitation in the relativistic theory is not more constraining than the maximal time limitation of the usual (non-relativistic) causality. Therefore, the inferred analyticity properties are quite similar. The main difference is that in the relativistic theory the dispersion relation is written in Lorentz-invariant variables, such as the Mandelstam variables s, t, and u. The non-relativistic dispersion relations must be cast in a Lorentz-invariant form. We shall see an example of that in chapter 3, where the Kramers–Kronig relation will be considered in some detail.

A much more comprehensive discussion of the relation between causality and analyticity can be found in, e.g. [4–7]. Let us pick up the argumentation from the point where it is established that the pertinent amplitude $f(z)$, for complex $z = \omega + i\gamma$, is an analytic function everywhere in the z plane except for the *branch cuts* along the real axis, such as those shown in figures 2.1 and 2.2. The branch cuts are physically associated with absorption and production mechanisms which usually have a threshold, indicated in the figures by ω_0.

The dispersion relation of type (1.1) comes about precisely because of these analytic properties. To see this, one needs to go through a bit of complex-variable analysis in the following section.

2.2 Derivation of dispersion relations

The starting point is the *Cauchy formula* for analytic functions:

$$f(z) = \frac{1}{2\pi i} \oint_C d\zeta \, \frac{f(\zeta)}{\zeta - z}, \tag{2.1}$$

where the integration is done along a closed contour C, which lies in the analyticity domain, i.e. anywhere except for the branch cut on the real axis. A convenient choice of C is shown in 2.1. The integral over C splits naturally into four terms,

$$f(z) = I_R(z) + I_+(z) + I_-(z) + I_r(z), \tag{2.2}$$

the integral I_R over the big circle of radius R, integrals $I_\pm$ above and below the branch cut, respectively, and integral I_r over the small semicircle of radius r around the threshold ω_0. Introducing $a = \arcsin(r/R)$, these integrals are written as follows:

[1] In principle, nothing forbids the motion of objects exceeding light speed relative to other distant objects. In fact, it is quite certain that some far away galaxies, and stars and other astronomical objects, move faster than light, with respect to us. They are then simply beyond our *causal horizon*, and hence, we cannot receive any signal from them, and hence, they cannot affect any phenomena we observe. Of course, these objects could mess with the causality by passing through the horizon, i.e. slowing down to subluminal speed with respect to us. Also, the questions arise of whether the superluminal objects could affect the observed objects gravitationally, or how they would fit in the cosmology of Universe, perhaps as a sort of dark matter. This is, however, a subject for a different story. From now on, causality is solid and whatever violates it (or whoever disagrees) is irrelevant.

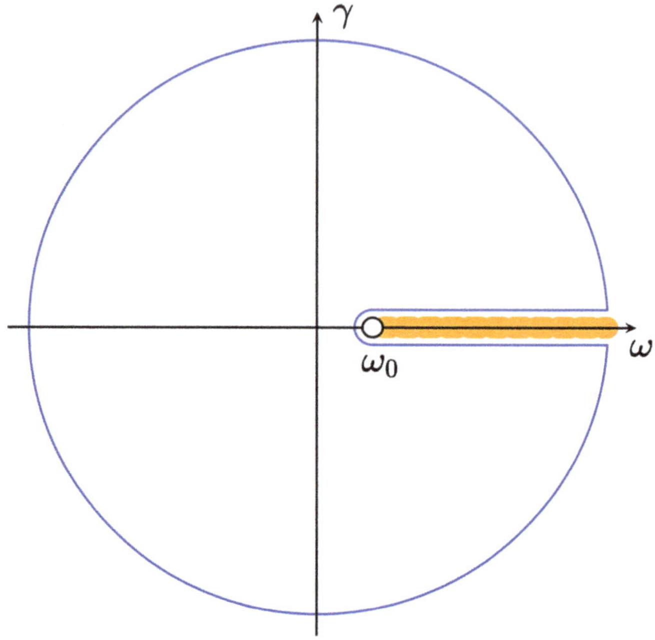

Figure 2.1. Analytic structure of a typical amplitude $f(z)$, with $z = \omega + i\gamma$, exhibiting a branch cut starting at ω_0. Enclosed in the contour is the domain of analyticity.

$$I_{\mathrm{R}}(z) = \frac{1}{2\pi} \int_{a}^{2\pi - a} d\varphi \; e^{i\varphi} \, \frac{f(Re^{i\varphi})}{e^{i\varphi} - z/R} \tag{2.3a}$$

$$I_{\pm}(z) = \pm \frac{1}{2\pi i} \int_{\omega_0}^{R} d\zeta \, \frac{f(\zeta \pm ir)}{\zeta \pm ir - z} \tag{2.3b}$$

$$I_{\mathrm{r}}(z) = -\frac{r}{2\pi} \int_{\pi/2}^{3\pi/2} d\varphi \; e^{i\varphi} \, \frac{f(\omega_0 + re^{i\varphi})}{\omega_0 + re^{i\varphi} - z}. \tag{2.3c}$$

Let us take the limits $r \to 0$ and $R \to \infty$. Then I_{r} vanishes, since $f(z)$ has no pole at ω_0. The condition for vanishing I_{R} is more tricky:

$$f(z) \xrightarrow{|z| \to \infty} 0. \tag{2.4}$$

On the other hand, it is the same condition which ensures that the integrals $I_{\pm}$ are convergent as $R \to \infty$. We will come back to the issue of convergence later on (see section 2.7).

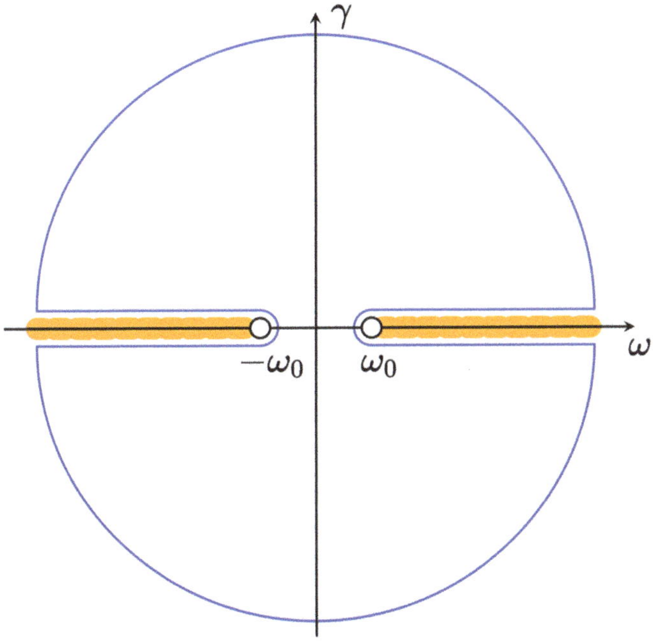

Figure 2.2. Analytic structure of a crossing invariant amplitude in the complex plane. The branch cuts are symmetric with respect to the imaginary axis.

With I_R and I_r vanishing, we are left with

$$f(z) = [I_+(z) + I_-(z)]_{r\to 0}^{R\to\infty}, \tag{2.5}$$

which can be simplified by means of the *Schwarz reflection principle* for analytic functions:

$$f^*(z) = f(z^*), \tag{2.6}$$

where the asterisk denotes the complex conjugation. Applying it to equation (2.3b) we find $I_-^*(z) = I_+(z^*)$, and eventually obtain:

$$f(z) = \lim_{r\to 0} \frac{1}{2\pi i} \int_{\omega_0}^{\infty} d\zeta \left[\frac{f(\zeta + ir)}{\zeta + ir - z} - \frac{f^*(\zeta + ir)}{\zeta - ir - z} \right] = \frac{1}{\pi} \int_{\omega_0}^{\infty} d\zeta \, \frac{\mathrm{Im} f(\zeta)}{\zeta - z}. \tag{2.7}$$

Thus, we can reconstruct the function in the entire complex plane from an integral of its imaginary part associated with the branch cut(s). Writing it for the real and imaginary parts separately, we have:

$$\mathrm{Re} f(\omega + i\gamma) = \frac{1}{\pi} \int_{\omega_0}^{\infty} d\zeta \, \frac{(\zeta - \omega)\, \mathrm{Im} f(\zeta)}{(\zeta - \omega)^2 + \gamma^2} \tag{2.8a}$$

$$\mathrm{Im}\,f(\omega + i\gamma) = \frac{1}{\pi}\int\limits_{\omega_0}^{\infty} d\zeta\, \frac{\gamma\,\mathrm{Im}\,f(\zeta)}{(\zeta - \omega)^2 + \gamma^2}. \tag{2.8b}$$

The dispersion relation (1.1) is obtained from equation (2.8a) by taking the limit to the real axis ($\gamma \to 0$):[2]

$$\mathrm{Re}\,f(\omega) = \lim_{\gamma\to 0}\frac{1}{\pi}\int\limits_{\omega_0}^{\infty} d\zeta\, \frac{(\zeta - \omega)\,\mathrm{Im}\,f(\zeta)}{(\zeta - \omega)^2 + \gamma^2} = \frac{1}{\pi}\fint\limits_{\omega_0}^{\infty} d\zeta\, \frac{\mathrm{Im}\,f(\zeta)}{\zeta - \omega}, \tag{2.9}$$

where the latter relation serves as a definition of the principal-value integration. Despite the different lower-bound of integration, this expression is identical to (1.1) because, for $\omega < \omega_0, \mathrm{Im}\,f(\omega) = 0$. Note also that the relation (2.8b), in the limit of $\gamma \to 0$, is satisfied identically by using the following definition of the Dirac's δ-function,

$$\delta(x) = \lim_{\gamma\to 0}\frac{\gamma}{x^2 + \gamma^2}, \tag{2.10}$$

and the well-known property $\int dx\, \delta(x - a) f(x) = f(a)$, for a within the integration interval.

2.2.1 An elementary example: the inverse square root

As a simple illustration of how the dispersion relation works, consider the following function:

$$f(\omega) = \frac{1}{\sqrt{\omega_0 - \omega}}. \tag{2.11}$$

It has the required analytic properties, and satisfies the convergence condition (2.4). Above the threshold, $\omega > \omega_0$, there is the branch cut and the corresponding imaginary part can be found as follows:[3]

$$\mathrm{Im}\,\frac{1}{(\omega_0 - \omega)^{\frac{1}{2}}} = \mathrm{Im}\, e^{-\frac{1}{2}\log(\omega_0 - \omega)} = \frac{1}{(\omega - \omega_0)^{\frac{1}{2}}}\,\theta(\omega - \omega_0), \tag{2.12}$$

where $\theta(x)$ is the step function (equal to 0 for $x < 0$ and 1 for $x \geqslant 0$). Plugging this into the right-hand side of the dispersion relation (2.9), one should recover equation (2.11) on the left-hand side. It is easiest to verify this for the choice of $\omega_0 = 0$:

---

[2] As an exercise, derive the dispersion relation for z lying on the branch cut, i.e., $\omega > \omega_0$ and $\gamma = 0$ from the outset. In this case the left-hand-side of the Cauchy formula is 0, since z is outside the contour. The final result is nonetheless the same.

[3] The logarithm develops an imaginary part for the negative argument:

$$\log(x \pm i0^+) = [\log(-x) \pm i\pi]\theta(-x),$$

where 0^+ is an infinitesimal positive number. When choosing the sign of the imaginary part we recall that the branch cut should be approached from above. That is the choice we have made in equation (2.7).

$$\frac{1}{\sqrt{-\omega}} = \frac{1}{\pi}\int_0^\infty d\omega' \frac{1}{(\omega' - \omega)\sqrt{\omega}}. \qquad (2.13)$$

2.3 Crossing symmetry

Very often, the function $f(\omega)$ has a definite parity under the reflection of energy, meaning it is either an even or odd function of ω,

$$f(-\omega) = \pm f(\omega). \qquad (2.14)$$

It arises as a result of *crossing symmetry*. For example, when we consider forward scattering of two particles, the amplitude should be invariant under an interchange of the incident and outgoing particles. The crossing symmetry for the examples of Compton and light-by-light scattering will be considered in chapters 4 and 6.

This property implies that the analytic structure is mirrored with respect to the imaginary axis, see figure 2.2. It is therefore convenient to do the above analysis in the complex plane of $z = \omega^2 + i\gamma$, rather then $z = \omega + i\gamma$. In the ω^2 complex plane, there is only one branch cut, just as we considered before, and the derivation follows through the same steps. When we arrive at the dispersion relation (2.9), we just need to change the variables from ω to ω^2, and analogously for the integration variable: $\zeta \to \zeta^2$. The resulting dispersion relations for the even or odd function are, respectively:

$$\mathrm{Re} f_{\mathrm{even}}(\omega) = \frac{2}{\pi}\fint_{\omega_0}^\infty d\zeta \frac{\zeta\,\mathrm{Im} f(\zeta)}{\zeta^2 - \omega^2}, \quad \mathrm{Re} f_{\mathrm{odd}}(\omega) = \frac{2\omega}{\pi}\fint_{\omega_0}^\infty d\zeta \frac{\mathrm{Im} f(\zeta)}{\zeta^2 - \omega^2}. \qquad (2.15)$$

2.4 Unitarity

Dispersion relations are most useful when the imaginary part is expressed in terms of observable quantities, usually photoabsorption cross sections. Unitarity, or conservation of probability, often facilitates such types of relations. For example, in the case of f describing the amplitude of forward Compton scattering off a target, unitarity results in the *optical theorem*:

$$\mathrm{Im} f(\omega) = \frac{\omega\,\sigma(\omega)}{4\pi}, \qquad (2.16)$$

with $\sigma(\omega)$ the total cross section of photoabsorption by that target. Combining this with crossing symmetry leads (for an even function) to:

$$\mathrm{Re} f(\omega^2) = \frac{1}{2\pi^2}\fint_{\omega_0}^\infty d\omega' \frac{\omega'^2\,\sigma(\omega')}{\omega'^2 - \omega^2}. \qquad (2.17)$$

This obviously is a very powerful result. Knowing the cross sections one can in principle evaluate the entire amplitude. The Kramers–Kronig relation for the refraction index (see chapter 3) and the Gell-Mann–Goldberger–Thirring relations for Compton scattering amplitudes (see chapter 4) are relations of this type. Similar relations exist as well for the light-by-light scattering amplitudes (see chapter 3). The relations for the vacuum polarization and electromagnetic form factors are also relations of this type, albeit without the use of the crossing.

2.5 Low-energy theorems and sum rules

The last step to turning the dispersion relations into sum rules is the low-energy (or, low-frequency) expansion. Expanding both sides of the dispersion relation (2.9) in a Taylor series around 0, and equating the coefficients, one obtains the following relations:

$$\begin{aligned} f(0) &= \frac{1}{\pi}\int_{\omega_0}^{\infty} d\omega\, \frac{\operatorname{Im} f(\omega)}{\omega}, \\ f'(0) &= \frac{1}{\pi}\int_{\omega_0}^{\infty} d\omega\, \frac{\operatorname{Im} f(\omega)}{\omega^2}, \\ &\vdots \\ f^{(n)}(0) &= \frac{n!}{\pi}\int_{\omega_0}^{\infty} d\omega\, \frac{\operatorname{Im} f(\omega)}{\omega^{n+1}}. \end{aligned} \tag{2.18}$$

Incidentally, if f is an even (or odd) function, we only need to pick, respectively, even (or odd) values of n and multiply the result by 2. This can be seen explicitly by expanding (2.15) by the power of ω.

When both sides of such a relation are expressed in terms of observable quantities, the relation becomes a *sum rule*. The right-hand side, as we have seen, is usually expressed in terms of a cross section. The left-hand side determines the low-energy behavior which often is fixed by the *low-energy theorems*.

2.5.1 The good, the bad, and the ugly?

Let us see how it works in the example of Compton scattering, i.e. an elastic scattering of a photon from a given target, whether it is an electron, a proton, or an entire atom. Omitting for now the spin complications (discussed in chapter 4), the amplitude of forward Compton scattering $f(\omega)$ depends on only the photon energy ω. As established above, the conditions of causality, crossing symmetry, and unitarity lead to the dispersion relation (2.17).

Now, the low-energy theorems (see e.g. [8, 9]) tell us that the first two terms in the energy expansion are as follows:

$$f(\omega) = -\frac{e^2}{4\pi M} + \left(\alpha_{\text{E1}} + \beta_{\text{M1}}\right)\omega^2 + O(\omega^4), \tag{2.19}$$

where the first one is the Thomson amplitude, given by the charge e and mass M of the target. The second term in the expansion is given by the dipole polarizabilities of the target, electric (α_{E1}) plus magnetic (β_{M1}). Expanding both sides of the dispersion relation we thus find the following two sum rules:

$$-\frac{e^2}{4\pi M} = \frac{1}{2\pi^2}\int_{\omega_0}^{\infty} d\omega\, \sigma(\omega), \tag{2.20}$$

$$\alpha_{\text{E1}} + \beta_{\text{M1}} = \frac{1}{2\pi^2}\int_{\omega_0}^{\infty} d\omega\, \frac{\sigma(\omega)}{\omega^2}. \tag{2.21}$$

The second sum rule here is the celebrated Baldin sum rule [10], which is proven to be very useful in data-driven assessments of the nucleon polarizabilities. We shall consider it in more detail in chapter 4. On the other hand, the first sum rule does not make much sense. This can never be correct, simply because the left-hand side is negative while the right-hand side is positive—the cross section is a positive-definite quantity.

Thus, we have obtained one good and one bad example of the sum rule. To fix the bad one, once and for all, we shall now re-examine our derivation of the dispersion relation of section 2.2.

2.6 Relaxing the convergence condition

In deriving the generic dispersion relation in section 2.2, we made a rather convenient, standard assumption about the asymptotic behavior of our function, see equation (2.4). What happens if we relax this condition? Obviously, we cannot drop the integral over the big semicircle I_{R}, whereas the integrals along the real axis $I_{\pm}$ will not converge (although their sum still might!).

Since I_{R} does not vanish we need to evaluate it starting from equation (2.3a). In the limit of $R \to \infty$, we have:

$$I_{R\to\infty} = \frac{1}{2\pi}\int_{0}^{2\pi} d\varphi \lim_{R\to\infty} f(Re^{i\varphi}). \tag{2.22}$$

Of help here will be the Sugawara–Kanazawa theorem [11], which states that the directional dependence of the function averages out and we can simply replace the integrand by the asymptotic value of the function, i.e.

$$\lim_{R\to\infty} f(Re^{i\varphi}) \to f(R\to\infty). \tag{2.23}$$

If the branch cut extends to infinity, one needs to take care that it is not crossed when evaluating I_{R}. In that case, only the real part of the function appears. Thus, the value of this integral is given by the real value of the function at infinite energy, which we denote as $f_\infty \equiv \mathrm{Re} f(\omega \to \infty)$.

Augmenting the resulting dispersion relation accordingly, we obtain

$$f(\omega) = f_\infty + \frac{1}{\pi} \,⨍_{\omega_0}^{\infty} d\zeta \, \frac{\mathrm{Im} f(\zeta)}{\zeta - \omega}. \tag{2.24}$$

With the extra term, the 'bad' sum rule (2.20) becomes

$$-\frac{e^2}{4\pi M} = f_\infty + \frac{1}{2\pi^2} \int_{\omega_0}^{\infty} d\omega \; \sigma(\omega). \tag{2.25}$$

The value of f_∞ will most likely be unknown, but at least the sum rule is not outright wrong. In fact it can be solved for f_∞ and substituted into the dispersion relation to obtain

$$\mathrm{Re} f(\omega^2) = -\frac{e^2}{4\pi M} + \frac{\omega^2}{2\pi^2} \,⨍_{\omega_0}^{\infty} d\omega' \, \frac{\sigma(\omega')}{\omega'^2 - \omega^2} \tag{2.26}$$

instead of (2.17). This is equivalent to a once-subtracted dispersion relation as considered in the following section. But before going there, we consider a simple example of the amplitude with divergent high-energy behavior.

2.6.1 An elementary example: the logarithm

As an illustration of the dispersion relation, corrected by the asymptotic term f_∞, consider the following function:

$$f(\omega) = \log(-\omega). \tag{2.27}$$

It has the needed analytical properties, the branch cut for positive ω, with

$$\mathrm{Im} f(\omega) = -\pi \, \theta(\omega),$$
$$f_\infty \equiv \mathrm{Re} f(\omega \to \infty) = \lim_{\omega \to \infty} \log \omega.$$

Substituting these into the right-hand side of the corrected relation (2.24), we have, in the limit of $\Lambda \to \infty$,

$$\log \Lambda + \frac{1}{\pi} \,⨍_{0}^{\Lambda} d\zeta \, \frac{-\pi}{\zeta - \omega} = \log \Lambda - \log \frac{\Lambda - \omega}{-\omega} \to \log(-\omega), \tag{2.28}$$

and hence the dispersion relation is verified, without obeying the convergence condition (2.4).

This example also provides a nice illustration of the Sugawara–Kanazawa theorem (see (2.22) and below),

$$I_{R\to\infty} = \frac{1}{2\pi}\int_0^{2\pi} d\varphi \lim_{R\to\infty} \log(-Re^{i\varphi}) = \lim_{R\to\infty} \log R + \frac{i}{2\pi}\int_0^{2\pi} d\varphi(\varphi - \pi). \qquad (2.29)$$

The last integral vanishes in accordance with the theorem.

2.7 Divergencies, subtractions, and renormalization

It is hard, if not impossible, to see *a priori* whether a given physical amplitude will satisfy the convergence condition or find out the value of f_∞. Both require knowledge of the amplitude at infinitely high energy. We normally do not have that information from experiments. We might know the behavior of the cross section at high energy, which will at least tell us whether the sum rule integral converges or not. For the proton, for example, the photoabsorption (γp) cross section increases at high energy, at least as far as one knows currently, see figure 2.3. The (off-forward) Compton scattering itself is one of the photoabsorption processes which will not cease at high energy.

Such a rising cross section compromises the unsubtracted dispersion relation (2.17), the integral diverges at infinity. As seen in the previous section, this divergence is supposed to be canceled by f_∞, the asymptotic value of the amplitude itself. It is better not to deal with this cancellation explicitly, we may consider the difference of the amplitude at two different energies, such that f_∞ drops out.

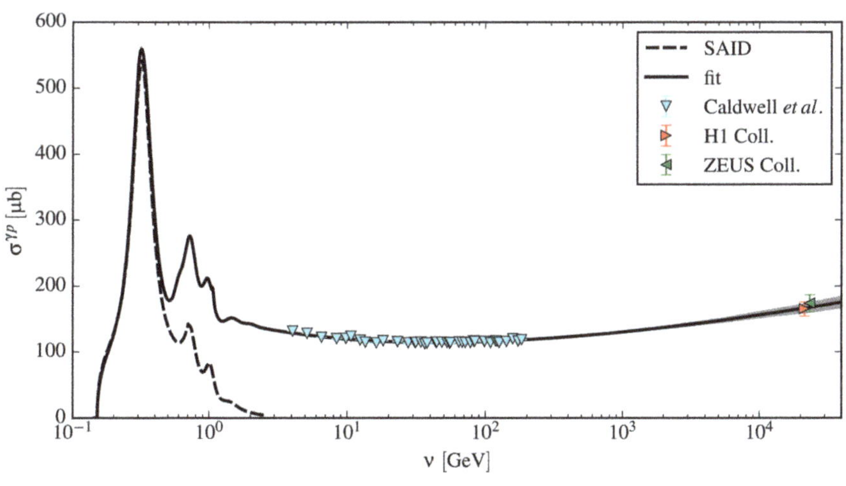

Figure 2.3. Total photoabsorption cross section for the proton as a function of photon energy in the proton rest frame. Dashed curve shows the contribution of single-pion production channels: $\pi^0 p$ and $\pi^+ n$. The legend for the experimental data is explained in figure 4.2. (Reproduced with permission from [12]. Copyright 2015 the American Physical Society.)

Continuing to work with the even amplitudes (the odd ones are analogous), let us write down a dispersion relation for the following function:

$$f_1(\omega^2) = \frac{f(\omega^2) - f(0)}{\omega^2}. \tag{2.30}$$

It obviously has a better asymptotic behavior than f, while preserving its analytic structure. At the same time, if we indeed know $f(0)$ from the low-energy theorem, we can always reconstruct f from f_1. Writing the analogue of equation (2.17) for f_1 and expressing equation (2.30) for f, we have:

$$\mathrm{Re} f(\omega^2) = f(0) + \frac{\omega^2}{2\pi^2} \fint_{\omega_0}^{\infty} d\omega' \, \frac{\sigma(\omega')}{\omega'^2 - \omega^2}. \tag{2.31}$$

The same result would be obtained if we subtract from each of the left- and right-hand sides of equation (2.17) their values at $\omega = 0$. This relation is, therefore, referred to as a *once-subtracted* dispersion relation.

The new relation is certainly an improvement: we have only given up determining $f(0)$ from the dispersion relations, the rest of the function is determined fine, provided the integral in the subtracted relation converges. If it does not, one can continue doing this trick, and so, for the n-times subtracted relation, we have:

$$\begin{aligned} \mathrm{Re} f(\omega^2) &= \sum_{k=1}^{n} f^{(k-1)}(0) \, \omega^{2k-2} + \frac{\omega^{2n}}{2\pi^2} \fint_{\omega_0}^{\infty} \frac{d\omega'}{\omega'^{2n-2}} \frac{\sigma(\omega')}{\omega'^2 - \omega^2}, \\ \text{with} \quad f^{(k)}(0) &= \left. \frac{d^k f(\omega^2)}{d\omega^{2k}} \right|_{\omega=0}, \end{aligned} \tag{2.32}$$

and the integer n can be chosen such that the integral converges. The price to pay is that we will need to determine n terms of the low-energy expansion of f from elsewhere, namely the low-energy theorems or other empirical information.

The physical interpretation of subtractions has a common ground with the renormalization program of quantum field theory (QFT). In many cases one can establish an exact correspondence between subtractions and the renormalization by casting the loop integrals into the form of dispersive integrals, e.g. by using the Cutkowski rules [13]. For example, the above subtraction of the Thomson term corresponds to the charge renormalization.

2.8 An approximate sum rule for the proton charge

The subtractions can be done at any other point, not only $\omega = 0$. In general, a subtraction at a fixed energy ω_1 casts equation (2.17) into the following dispersion relation:

$$\mathrm{Re} f(\omega^2) = \mathrm{Re} f(\omega_1^2) + \frac{(\omega^2 - \omega_1^2)}{2\pi^2} \fint_{\omega_0}^{\infty} d\omega' \frac{\sigma(\omega')}{\omega'^2 - \omega^2} \frac{\omega'^2}{\omega'^2 - \omega_1^2}. \tag{2.33}$$

Of course, this is only useful if $f(\omega_1^2)$ is known. Indeed, if we were to know it for the much-discussed case of Compton scattering, we would have a meaningful sum rule for the Thomson term:

$$-\frac{e^2}{4\pi M} = \mathrm{Re} f(\omega_1^2) - \frac{\omega_1^2}{2\pi^2} \fint_{\omega_0}^{\infty} d\omega' \frac{\sigma(\omega')}{\omega'^2 - \omega_1^2}. \tag{2.34}$$

In this respect, it is interesting to observe (in figure 2.4) that the Compton amplitude for the proton is vanishing near the first strong threshold, i.e. the threshold for pion production. The two curves in 2.4 show the real and imaginary parts of the spin-independent amplitude $f(\nu)$, which are determined from, respectively, the once-subtracted dispersion relation and the optical theorem upon inserting therein the empirical γp cross section shown in 2.3. One can see that the amplitude crosses 0 at a point slightly above the threshold for the process of charge-pion production ($\gamma p \to p^+ n$).

This could just be a coincidence, or there could be a deep reason for it, we do not know. But in any case, it warrants an approximate sum rule for the proton charge e:

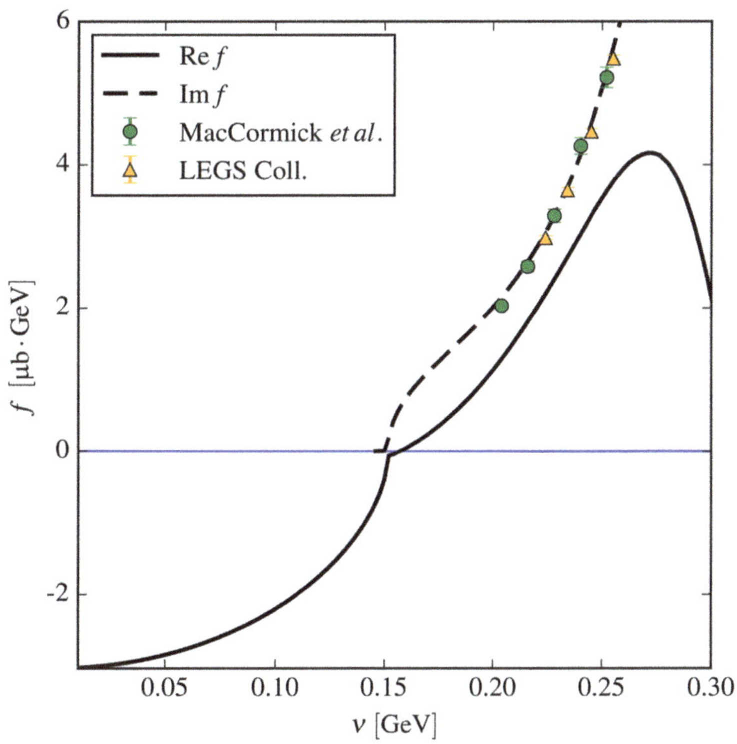

Figure 2.4. The spin-independent forward amplitude for Compton scattering off the proton resulting, by means of the dispersion relation, from the photoabsorption cross section of figure 2.3. Further details, such as references to experimental data, can be found in chapter 4. (Reproduced with permission from [12]. Copyright 2015 the American Physical Society.)

$$\frac{e^2}{2M_p} \approx \frac{\nu_\pi^2}{\pi} \int_{\nu_\pi}^{\infty} d\nu \, \frac{\sigma^{\gamma p}(\nu)}{\nu^2 - \nu_\pi^2}, \tag{2.35}$$

written here in the Lorentz-invariant variable $\nu = (s - M_p^2)/2M_p$, where s is the Mandelstam invariant and $M_p \simeq 0.938$ GeV is the proton mass. This variable corresponds with the photon energy in the lab frame (proton rest frame), with the pion-production threshold $\nu_\pi \simeq m_\pi + m_\pi^2/2M_p$ located at approximately 0.15 GeV; m_π is the pion mass.

The relation (2.35) is a peculiar example of sum rules. Unlike the others, it is a rigorous prediction based on general principles. It relies on an empirical fact (the vanishing amplitude in the vicinity of the threshold), which may or may not have a systematic foundation behind it. Until such a foundation is found, we cannot judge quantitatively how good this approximation is.

References

[1] Kronig R 1942 Algemeene Theorie der diëlectrische en magnetische Verliezen *Tijdschr. Ned. Radiogen.* **10** 2–9

[2] van Kampen N G 1953 S-matrix and causality condition. I. Maxwell field *Phys. Rev.* **89** 1072–9

[3] van Kampen N G 1953 S-matrix and causality condition. II. Nonrelativistic particles *Phys. Rev.* **91** 1267–76

[4] Gell-Mann M, Goldberger M and Thirring W E 1954 Use of causality conditions in quantum theory *Phys. Rev.* **95** 1612–27

[5] Toll J S 1956 Causality and the dispersion relation: logical foundations *Phys. Rev.* **104** 1760–70

[6] Screaton G R 1961 *Dispersion Relations* (Edinburgh: Oliver and Boyd)

[7] Nussenzveig H M 1972 *Causality and Dispersion Relations, Mathematics in Science and Engineering* (London: Elsevier Science)

[8] Low F E 1954 Scattering of light of very low frequency by systems of spin 1/2 *Phys. Rev.* **96** 1428–32

[9] Gell-Mann M and Goldberger M L 1954 Scattering of low-energy photons by particles of spin 1/2 *Phys. Rev.* **96** 1433–8

[10] Baldin A M 1960 Polarizability of nucleons *Nucl. Phys.* **18** 310–7

[11] Sugawara M and Kanazawa A 1961 Subtractions in dispersion relations *Phys. Rev.* **123** 1895–902

[12] Gryniuk O, Hagelstein F and Pascalutsa V 2015 Evaluation of the forward Compton scattering off protons: spin-independent amplitude *Phys. Rev.* D***92*** *074031*

[13] Cutkosky R E 1960 Singularities and discontinuities of Feynman amplitudes *J. Math. Phys.* **1** 429–33

Chapter 3

The Kramers–Kronig relation

The first and best-known example of dispersion relations is the Kramers–Kronig relation for the refraction of light in a dilute medium,

$$n(\omega) - 1 = \frac{N}{\pi} \fint_{\omega_0}^{\infty} d\omega' \, \frac{\sigma(\omega')}{\omega'^2 - \omega^2}. \tag{3.1}$$

Here $n(\omega) = c/v(\omega)$ is the refraction index, i.e. the ratio of the speed of light in vacuum and the medium. On the right-hand side we have: N, the number density (per unit volume) of the medium constituents, $\sigma(\omega)$, their total photo-absorption cross section, and ω_0, the threshold of the photo-absorption process. In what follows we use the *natural units* in which $c = 1$ and ω is the photon energy.

The above expression neglects the constituents' motion, which is not a bad approximation for non-relativistic systems. It is instructive, however, to see how the relation works with the medium being a gas of relativistic particles of mass M, characterized by a momentum-dependent number density $\mathcal{N}(\vec{p}\,)$. Later, we can take the limit of $M \to \infty$ to recover the non-relativistic formulae, or $M \to 0$ to find the ultra-relativistic result. The latter case will then be applied to the refraction index of the cosmic microwave background (CMB).

3.1 Refraction in a relativistic medium

Let us see how the above Kramers–Kronig relation generalizes to a dilute medium consisting of moving particles. In a Lorentz-invariant theory, the cross section is a function of the invariant energy $s = (P + Q)^2$, with P the four-momentum of the medium constituent absorbing the photon with four-momentum Q. For a photon with energy ω moving along the z-axis, the four-momentum is given by $Q = (\omega, 0, 0, \omega)$, whereas the constituent particle absorbing the photon has in general the four-momentum $P = (\varepsilon_p, \vec{p}\,)$, where $\varepsilon_p = \sqrt{M^2 + \vec{p}^{\,2}}$ is the energy. Hence,

$$s = P^2 + 2P \cdot Q + Q^2 = M^2 + 2\omega(\varepsilon_p - p \cos \vartheta), \tag{3.2}$$

doi:10.1088/978-0-7503-3431-0ch3

with $p = |\vec{p}|$ being the momentum and ϑ the angle at which the constituent encounters the photon. It is useful to introduce another kinematical invariant:

$$\nu = \frac{P \cdot Q}{M} = \frac{s - M^2}{2M} = \omega\gamma(1 - \beta\cos\vartheta), \tag{3.3}$$

where $\beta = p/\varepsilon_\text{p}$ is the velocity and $\gamma = \varepsilon_\text{p}/M = 1/\sqrt{1 - \beta^2}$ is the Lorentz-contraction factor. In the rest frame of the constituent, or in the non-relativistic limit $\beta \to 0$ (equivalently, $M \to \infty$), the invariant ν has the interpretation of the photon energy, i.e. is identical to ω. We, however, assume that only the whole system is at rest, whereas the constituents are moving with a certain momentum distribution. Thus, ν and ω are generally different, and since the threshold in the variable ν is a Lorentz-invariant constant ν_0, the threshold in ω depends on the constituent momentum. Namely, from (3.3), the lowest value of ω is

$$\omega_0 = \frac{M\nu_0}{\varepsilon_\text{p} + p} = \nu_0\sqrt{\frac{1 - \beta}{1 + \beta}}. \tag{3.4}$$

With these definitions, the Kramers–Kronig relation takes the following form[1]

$$n(\omega) - 1 = \int \frac{d\vec{p}}{(2\pi)^3}\,\mathcal{N}(\vec{p})\,\frac{1}{\pi}\fint_{\omega_0}^{\infty} d\omega' \frac{\sigma(\nu')}{\omega'^2 - \omega^2}, \tag{3.5}$$

where the momentum distribution is normalized to the number density

$$\int \frac{d\vec{p}}{(2\pi)^3}\,\mathcal{N}(\vec{p}) = N, \tag{3.6}$$

and the cross section is now involved in the momentum integration, because the relation of ν' and ω' involves $\vec{p}$ as follows: $\nu' = (\omega'/M)(\varepsilon_\text{p} - p\cos\vartheta)$.

A truly Lorentz-invariant result can be obtained for a spherically symmetric momentum distribution, $\mathcal{N}(\vec{p}) = \mathcal{N}(p)$. In this case, the directions of $\vec{p}$ can be integrated out. Integration over the azimuthal direction yields 2π and, introducing $z = \cos\vartheta$, we are left with

---

[1] This expression has an annoying caveat. The threshold mismatch pointed out in (3.4) seems to constrain the integration over $\cos\vartheta$ contained in the three-dimensional integration over $\vec{p}$. Normally, the integration over the polar angle ϑ goes from 0 to π, or from -1 to 1 over $z = \cos\vartheta$. However, since $\nu' = \omega'\gamma(1 - \beta z) \geqslant \nu_0$, one finds

$$\cos\vartheta \leqslant \frac{1}{\beta}\left(1 - \frac{\nu_0}{\gamma\omega'}\right) = \frac{1}{\beta}\left[1 - (1 + \beta)\frac{\omega_0}{\omega'}\right].$$

The right-hand side of this inequality is equal to -1 at the threshold, $\omega' = \omega_0$, and then increases with ω', reaching 1 at $\omega' = \omega_0(1 + \beta)/(1 - \beta)$. Rather than implementing this cumbersome constraint in the integration limits, we set the lower limit of integration in energy to 0, i.e. setting both ν_0 and ω_0 to 0 in the above equations. It is understood that the threshold is set by a step function, $\theta(\nu - \nu_0)$, contained in the cross section. The integration over $\cos\vartheta$ is then not additionally constrained—problem solved!

$$n(\omega) - 1 = \frac{1}{(2\pi)^2} \int_0^\infty dp\, p^2\, \mathcal{N}(p) \int_{-1}^1 dz\, \frac{1}{\pi} \fint_0^\infty d\omega' \frac{\sigma(\nu')}{\omega'^2 - \omega^2}. \tag{3.7}$$

To make the z dependence explicit, we change the integration variables ω' to ν', such that

$$\frac{d\omega'}{\omega'^2 - \omega^2} = \frac{d\nu'}{\omega^2} \frac{\gamma(1 - \beta z)}{(\nu'/\omega)^2 - \gamma^2(1 - \beta z)^2}. \tag{3.8}$$

We may then interchange the order of z and ν' integrations, and perform the polar-angle integration as follows (for $a = \nu'/\omega$):

$$\begin{aligned}\mathcal{L}(a^2) &\equiv \frac{1}{2} \fint_{-1}^1 dz\, \frac{\gamma(1 - \beta z)}{a^2 - \gamma^2(1 - \beta z)^2} \\ &= \frac{1}{4} \fint_{-1}^1 dz \left[\frac{1}{a - \gamma(1 - \beta z)} - \frac{1}{a + \gamma(1 - \beta z)}\right] \\ &= \frac{1}{4\beta\gamma} \log \left| \frac{a - \gamma(1 - \beta)}{a - \gamma(1 + \beta)} \cdot \frac{a + \gamma(1 - \beta)}{a + \gamma(1 + \beta)} \right| = \frac{1}{4\beta\gamma} \log \left| \frac{a^2 - \gamma^2(1 - \beta)^2}{a^2 - \gamma^2(1 + \beta)^2} \right|.\end{aligned} \tag{3.9}$$

Note that the principal-value integration has, while switching the order of integrations, been switching too. It eventually results in the absolute-value operation on the logarithm's argument.

We finally arrive at the following expression:

$$n(\omega) - 1 = \frac{4\pi}{(2\pi)^3\omega^2} \int_0^\infty dp\, p^2\, \mathcal{N}(p)\, \frac{1}{\pi} \int_{\nu_0}^\infty d\nu' \sigma(\nu')\, \mathcal{L}\left(\frac{\nu'^2}{\omega^2}\right), \tag{3.10}$$

which can be considered as a relativistic generalization of the Kramers–Kronig relation. Incidentally, for small velocities, i.e. the non-relativistic limit ($\beta \approx 0$),

$$\frac{1}{\omega^2} \mathcal{L}\left(\frac{\nu'^2}{\omega^2}\right) = \frac{1}{\nu'^2 - \omega^2} + O(\beta^2), \tag{3.11}$$

and using the equivalence of ν' and ω' in this limit, and the normalization (3.6), we reproduce the standard expression (3.1). Our effort to relativize the Kramers–Kronig relation pays off, when considering the *ultra-relativistic* limit, $\beta \to 1$. The constituents are then moving with speeds approaching the speed of light, implying $\varepsilon_{\rm p} = p$, and hence their mass is negligible.

The massless limit ($M \to 0$) can be taken simply in the relativized relation (3.10), one just needs to be careful. First of all, recall that the definition of the invariant photon energy ν involves the mass, i.e. $\nu = (s - M^2)/2M$. To make that explicit, we change the integration from ν' to s'. Second, in the kernel function (3.9) we substitute $\gamma = \varepsilon_{\rm p}/M$. After these steps, the massless limit can be taken in a straightforward manner:

$$n(\omega) - 1 = \frac{4\pi}{(2\pi)^3\omega^2}\int_0^\infty dp\, p^2\, \mathcal{N}(p)\, \frac{1}{\pi}\int_{s_0}^\infty \frac{ds'}{2M}\,\sigma(s')\,\frac{M}{4\beta\varepsilon_p}\log\left|\frac{\frac{(s'-M^2)^2}{\omega^2} - 4\varepsilon_p^2(1-\beta)^2}{\frac{(s'-M^2)^2}{\omega^2} - 4\varepsilon_p^2(1+\beta)^2}\right| \tag{3.12}$$

$$\overset{M\to 0}{=} -\frac{4\pi}{(2\pi)^3\omega^2}\int_0^\infty dp\, p\, \mathcal{N}(p)\, \frac{1}{8\pi}\int_{s_0}^\infty ds'\sigma(s')\log\left|1 - \frac{16p^2\omega^2}{s'^2}\right|.$$

The latter expression is therefore the ultra-relativistic form of the Kramers–Kronig relation, which applies, most notably, to the case when the medium is a photon gas, such as the CMB radiation. Before going into that in detail (section 3.3), let us point out a few general properties of the Kramers–Kronig relation and see, in particular, how the sum rules appear in this picture.

3.2 The low-frequency limit: the Lorentz–Lorenz relation

The connection with the sum rules becomes apparent for low frequencies ($\omega \ll \nu_0$), when the refraction index can be expanded in powers of ω. Consider the low-energy expansion of the relativized relation (3.10). The Taylor-series expansion of the kernel function (3.9) for small ω yields

$$\mathcal{L}\left(\frac{\nu'^2}{\omega^2}\right) = \frac{\omega^2}{\nu'^2}\gamma + \frac{\omega^4}{\nu'^4}\gamma^3(1+\beta^2) + \cdots \tag{3.13}$$

and substituting this into (3.10), we see that the two integrations factorize, and the integrals over ν' have the form

$$I_k = \frac{2}{\pi}\int_{\nu_0}^\infty d\nu'\,\frac{\sigma(\nu')}{\nu'^k}, \tag{3.14}$$

with even k. These are precisely the integrals which arise in the sum rules for Compton scattering, see chapter 4. For instance, the Baldin sum rule states that $I_2 = 4\pi(\alpha_{E1} + \beta_{M1})$, i.e. the sum of the dipole electric and magnetic polarizabilities. Hence, for the first term in the expansion we obtain

$$n(0) - 1 = 2\pi(\alpha_{E1} + \beta_{M1})\,\frac{\varrho}{M}, \tag{3.15}$$

with the relativistic energy density

$$\varrho = \frac{4\pi}{(2\pi)^3}\int_0^\infty dp\, p^2\,\sqrt{M^2+p^2}\,\mathcal{N}(p). \tag{3.16}$$

The latter quantity is, in the non-relativistic limit, given by MN, and furthermore neglecting the magnetic polarizability we recover the approximate *Lorentz–Lorenz relation*:

$$n(0) - 1 \approx 2\pi N\alpha_{E1}. \tag{3.17}$$

This is also what we would obtain by taking the low-frequency limit ($\omega \to 0$) directly in the standard expression (3.1). The result of (3.15) is, however, more accurate.

An analogous result can be obtained in the ultra-relativistic limit, even though the notion of polarizabilities is less of an obvious concept for a massless particle. For instance, in case of the photon gas considered further below, the analogue of (3.15) is (3.23), where the role of polarizabilities is played by the structure constants describing the low-energy photon self-couplings, i.e. non-linear electromagnetic effects.

3.3 CMB refraction index

The ultimate ultra-relativistic medium, capable of refracting and dispersing light, is the CMB radiation, consisting of light itself[2]. As we shall see from the following calculation, its actual effect on cosmic γ-rays, in the presently observed energy range, should be very tiny, and practically undetectable. This case is therefore considered here for 'academic' purposes only.

The starting point is our ultra-relativistic formula, (3.12), with the photo-absorption cross section given by the so-called total cross section of $\gamma\gamma$-fusion, $\sigma^{\gamma\gamma}(s)$. The CMB is an ideal photon gas with temperature $T \simeq 2.725$ K (and slowly getting colder, due to the inflation). Hence, its number density distribution in energy ε is given by the Bose–Einstein distribution, $\mathcal{N}(\varepsilon) = 1/(e^{\varepsilon/T} - 1)$. Sticking it into (3.12), we have

$$n_{\mathrm{CMB}}(\omega) - 1 = -\frac{4\pi}{(2\pi)^3\omega^2}\int_0^\infty d\varepsilon\,\frac{\varepsilon}{e^{\varepsilon/T}-1}\,\frac{1}{8\pi}\int_{s_0}^\infty ds\,\sigma^{\gamma\gamma}(s)\log\left|1-\frac{(4\varepsilon\omega)^2}{s^2}\right|, \tag{3.18a}$$

$$= \frac{\varrho_\gamma}{\pi}\int_{s_0}^\infty ds\,\frac{\sigma^{\gamma\gamma}(s)}{s^2} + \frac{160\pi^2\varrho_\gamma}{21}T^2\omega^2\frac{2}{\pi}\int_{s_0}^\infty ds\,\frac{\sigma^{\gamma\gamma}(s)}{s^4} + O(\omega^4), \tag{3.18b}$$

where

$$\varrho_\gamma \equiv \int_0^\infty \frac{d\varepsilon}{\pi^2}\,\frac{\varepsilon^3}{e^{\varepsilon/T}-1} = \frac{\pi^2}{15}T^4 \approx 3\times 10^{-38}\ \mathrm{MeV}^4 \tag{3.19}$$

is the CMB energy density. The remaining integrals of the $\gamma\gamma$-fusion cross section, appearing in the low-frequency expansion, must be involved in some sum rules. The first integral is given by the Euler–Heisenberg constants of the low-energy photon self-interactions (see chapter 6):

$$c_1 + c_2 = \frac{1}{4\pi}\int_{s_0}^\infty ds\,\frac{\sigma^{\gamma\gamma}(s)}{s^2}. \tag{3.20}$$

In the Standard Model, the dominant mechanism for the $\gamma\gamma$ fusion is the e^+e^- pair production, given to leading order in α as (see figure 3.1)

[2] Yes, light interacts with light. If you, the wander-full reader, find it mind-boggling, then welcome to the club! But only if you have recently joined a club. Otherwise, keep reading… Not this book though—it is either too complicated or too simple for you.

Figure 3.1. Leading order diagrams of $\gamma\gamma \to e^+e^-$.

$$\sigma^{(\gamma\gamma\to e^+e^-)} = \frac{4\pi\alpha^2}{s}\left\{2\left(1+\frac{4m_e^2}{s}-\frac{8m_e^4}{s^2}\right)\operatorname{arccosh}\frac{\sqrt{s}}{2m_e}-\left(1+\frac{4m_e^2}{s}\right)\sqrt{1-\frac{4m_e^2}{s}}\right\}, \quad (3.21)$$

with $m_e \simeq 0.511$ MeV the electron mass, $\alpha = e^2/4\pi \simeq 1/137$ the fine-structure constant. Substituting this cross section into the low-frequency expansion of the refraction index (3.18b) with $s_0 = 4m_e^2$, we obtain

$$n_{\rm CMB}(\omega) - 1 = \frac{11\alpha^2\varrho_\gamma}{90m_e^4} + \frac{104\alpha^2\varrho_\gamma}{2835m_e^8}\pi^2T^2\omega^2 + O(\omega^4, \alpha^4). \quad (3.22)$$

One can see that the CMB refraction index is astronomically small: $n_(0) - 1 \approx 3 \times 10^{-42}$.

In general, however, the exact expression is

$$n_{\rm CMB}(0) - 1 = 4(c_1 + c_2)\varrho_\gamma. \quad (3.23)$$

The above numerical results only correspond with the values of the low-energy constants in leading-order quantum electrodynamics (QED): $c_1 = 1/90\,\alpha^2 m_e^{-4}$, $c_2 = 7/360\,\alpha^2 m_e^{-4}$. Until these constants are measured directly, there is room for new physics scenarios (axions, dark matter, supersymmetry, etc), which could be explored using this sort of calculation.

IOP Publishing

Causality Rules (Second Edition)
Dispersion theory in non-elementary particle physics
Vladimir Pascalutsa

Chapter 4

Sum rules for Compton scattering

The general principles of causality and unitarity, applied to Compton scattering, result in a number of remarkable relations, *viz*, sum rules. Here we go through their derivation and consider a few prominent examples and applications.

4.1 Forward kinematics: helicity amplitudes for any spin

Consider the elastic scattering of a photon—Compton scattering—off a massive particle. We limit ourselves to the forward kinematics, where the four-momenta of the particles before and after scattering are the same, see figure 4.1. The Mandelstam invariant t, which represents the amount of transferred momentum, therefore vanishes ($t = 0$). The other two invariants are

$$s = (p + q)^2 = M^2 + 2M\nu - Q^2, \tag{4.1a}$$

$$u = (p - q)^2 = M^2 - 2M\nu - Q^2, \tag{4.1b}$$

where M is the mass of the target particle, and $Q^2 = -q^2$ is the space-like photon virtuality. The invariant

$$\nu = \frac{p \cdot q}{M} = \frac{s - u}{4M} \tag{4.2}$$

has the interpretation of the photon energy in the lab frame (i.e. the rest frame of the target). It has an important property of reflection $\nu \to -\nu$ under crossing $s \leftrightarrow u$. Throughout this chapter we assume $Q^2 = 0$, i.e. the photon is real. The case of virtual photons is considerably more involved and will be studied separately in chapter 5.

To deal with the spin of both the photon and the target, it is convenient to use the helicity formalism (helicity being the projection of a particle's spin on its

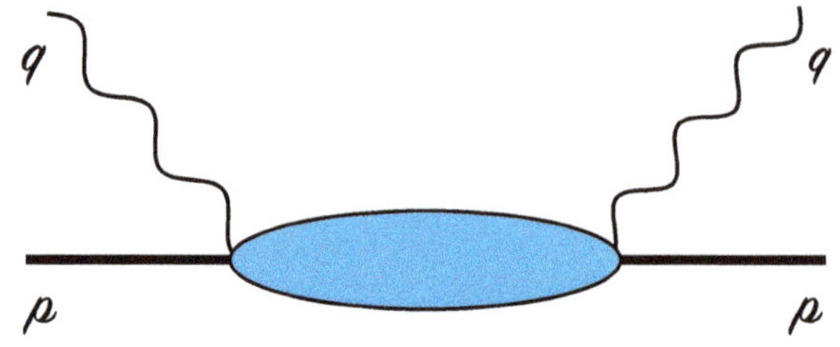

Figure 4.1. Forward Compton scattering, with q the four-momentum of the photon and p the four-momentum of the target particle.

momentum). The entire process of Compton scattering, at $t = 0$, can be described by helicity amplitudes, $T_{\lambda' h' \lambda h}(\nu)$, which, besides the invariant energy ν, depend on the helicities of the initial and final photon, λ and λ', and the corresponding helicities of the target particle, h and h'.

A distinct property of forward scattering, which follows from the conservation of total angular momentum (as there is no orbital momentum in forward scattering), is the conservation of total helicity, i.e. $H = H'$, with[1] $H = \lambda - h$ and $H' = \lambda' - h'$. In the case of a spinless target, this simply means $T_{\lambda'\lambda}(\nu) = \delta_{\lambda'\lambda}\, T(\nu)$, and hence everything is described by a single scalar amplitude.

In general, for a target with spin S, the forward amplitude depends on the total helicity H, running from $-(S + 1)$ to $S + 1$, as well as on the change of the photon helicity $\Delta\lambda = \lambda - \lambda'$. For a real photon, the possible helicities are ± 1, and hence $\Delta\lambda = -2, 0, 2$. Note that any non-vanishing $\Delta\lambda$ has to be compensated by a change of the target helicity, such that H stays constant. Therefore, to have $\Delta\lambda \neq 0$, the target must have at least spin one ($S \geqslant 1$), and the magnitude of total helicity must be less than maximal by at least 2 units ($|H| \leqslant S - 1$). Thus, so far we have $2S + 3$ amplitudes with $\Delta\lambda = 0$, and $(2S - 1) \times 2$ amplitudes with $\Delta\lambda \neq 0$.

Furthermore, the usual constraints on the helicity dependence from discrete symmetries—parity (P) and time-reversal (T)—applied to the forward amplitude, read

$$\text{P:} \quad T_{\mathrm{H},\Delta\lambda} = T_{-\mathrm{H},-\Delta\lambda}, \qquad \text{T:} \quad T_{\mathrm{H},\Delta\lambda} = T_{\mathrm{H},-\Delta\lambda} \qquad \text{PT:} \quad T_{\mathrm{H},\Delta\lambda} = T_{-\mathrm{H},\Delta\lambda}. \tag{4.3}$$

This means the amplitude does not depend on the sign of either H or $\Delta\lambda$. To count the number of amplitudes reduced by this constraint we now need to distinguish whether S is integer (bosons) or half-integer (fermions). In the latter case the number of independent H values is cut by two, while in the former special care must be taken of $H = 0$ in addition. Thus, the number of independent amplitudes, for either a boson or a fermion target, counts as follows:

[1] The minus sign in adding helicities appears here because of the convention that the photon is moving in the opposite direction.

$$\text{Boson:} \quad \frac{1}{2}(2S+2)+1+\frac{1}{2}(2S-2)+1=2S+2, \tag{4.4}$$

$$\text{Fermion:} \quad \frac{1}{2}(2S+3)+\frac{1}{2}(2S-1)=2S+1. \tag{4.5}$$

To recapitulate, this counting (of independent amplitudes) applies to the forward Compton scattering off a massive target with non-vanishing spin S.

Note that the number of amplitudes is always even (sparing the spinless and massless cases). Let's illustrate this with a few simple examples[2].

For $S = 1/2$ there are two independent amplitudes, e.g. $T_{1/2,\,0} = T_{1\,1/2\,1\,1/2}$ and $T_{3/2,\,0} = T_{1\,-1/2\,1\,-1/2}$.

For $S = 1$ there are four: $T_{0,\,0} = T_{1\,1\,1\,1}$, $T_{1,\,0} = T_{1\,0\,1\,0}$, $T_{2,\,0} = T_{1\,-1\,1\,-1}$, and $T_{0,\,2} = T_{1\,1\,-1\,-1}$.

For $S = 3/2$ there are also four: $T_{1/2,\,0}$, $T_{3/2,\,0}$ as for $S = 1/2$, and $T_{5/2,\,0} = T_{1\,-3/2\,1\,-3/2}$, $T_{1/2,\,2} = T_{1\,1/2\,-1\,-3/2}$.

In the case of virtual photons, one needs in addition to consider the amplitudes involving the longitudinal photon polarization. I leave it up the reader to prove that the number of extra amplitudes is $3S + 1$ for a boson and $3S + 1/2$ for a fermion.

The next step is to relate these amplitudes to the various polarized photoabsorption cross sections.

4.2 Optical theorem: dispersion relation

The choice of the forward kinematics was preconditioned of course by our wily plan to express everything in terms of the photoabsorption cross sections. This plan is realized in two major steps: (i) the optical theorem, which expresses the imaginary part of the amplitude in terms of the cross sections; and (ii) the dispersion relation, which expresses the real part as an integral of the imaginary part.

The spinless case is easy and we have started to consider it in chapter 2. Let us look at it in more detail. The forward Compton scattering amplitude for a spinless target is denoted as $f(\nu)$, where ν is the invariant photon energy defined in (4.2). It really is the photon energy in the laboratory frame where the target is at rest, $p = (M, \vec{0})$. Recall the optical theorem:

$$\mathrm{Im} f(\nu) = \frac{\nu}{4\pi}\sigma(\nu), \tag{4.6}$$

where σ is the total photoabsorption cross section. The unsubtracted dispersion relation, including the crossing symmetry, reads (omitting for now the possible complication due to $f(\infty) \neq 0$):

---

[2] Here the right-hand side is written in the standard notation $T_{\lambda' h' \lambda h}$, whereas the left-hand side is written as $T_{\mathrm{H},\Delta\lambda}$, with $H = \lambda - h = \lambda' - h'$ and $\Delta\lambda = \lambda - \lambda'$.

$$\mathrm{Re} f(\nu) = \frac{1}{2\pi^2} \fint_0^\infty d\nu' \, \frac{\nu'^2 \, \sigma(\nu')}{\nu'^2 - \nu^2}. \tag{4.7}$$

In what follows, we may often combine these two statements, (4.6) and (4.7), into one, by introducing an infinitesimal positive number 0^+ and writing

$$f(\nu) = \frac{1}{2\pi^2} \int_0^\infty d\nu' \, \frac{\nu'^2 \, \sigma(\nu')}{\nu'^2 - \nu^2 - i0^+}. \tag{4.8}$$

For a target with spin, there is one amplitude which can be written in exactly the same way—the spin-averaged (or, spin-independent) amplitude:

$$f(\nu) = \frac{1}{2(2S+1)} \sum_{\lambda=\pm 1} \sum_{h=-S}^{S} T_{\lambda h \lambda h}(\nu) = \frac{1}{2S+1} \sum_{h=-S}^{S} T_{|1-h|,0}, \tag{4.9}$$

which shares the optical theorem with the unpolarized cross section of total photoabsorption.

For the helicity amplitudes with no flip of photon helicity ($\Delta\lambda = 0$), the optical theorem is straightforward as well:

$$\mathrm{Im}\, T_{\mathrm{H},0}(\nu) = \frac{\nu}{4\pi} \sigma_{\mathrm{H}}(\nu), \tag{4.10}$$

where $\sigma_{\mathrm{H}}(s)$ is the doubly polarized cross section with circularly polarized photons. For example, in the case of $S = 1/2$, besides the spin-independent amplitude, $f(\nu) \equiv \frac{1}{2}[T_{1/2,\,0}(\nu) + T_{3/2,\,0}(\nu)]$, one may define the spin-dependent amplitude:

$$g(\nu) \equiv \frac{1}{2}[T_{1/2,\,0}(\nu) - T_{3/2,\,0}(\nu)], \tag{4.11}$$

which, using the fact that it must be odd under crossing, i.e. $g(-\nu) = -g(\nu)$, is expressed as follows:

$$\boxed{g(\nu) = \frac{\nu}{4\pi^2} \int_0^\infty d\nu' \, \frac{\nu'[\sigma_{1/2}(\nu') - \sigma_{3/2}(\nu')]}{\nu'^2 - \nu^2 - i0^+}.} \tag{4.12}$$

For the amplitudes with a photon helicity flip ($\Delta\lambda = 2$) the situation is more tricky, and involves the doubly polarized cross sections with linear (rather than circular) photon polarizations. For example, in the $S = 1$ case, where such an amplitude first appears, the optical theorem reads

$$\mathrm{Im}\, T_{0,\,2}(\nu) = \frac{\nu}{4\pi} \left[\sigma_{||}(\nu) - \sigma_{\perp}(\nu)\right], \tag{4.13}$$

where $\sigma_{||}$ (or $\sigma_{\perp}$) stands for the photoabsorption cross section with the photon linearly polarized along (or perpendicular to) the transverse polarization of the target. The closest we will come considering this situation will be in chapter 6, where the role of the spin-1 target is played by another photon. In the rest of this chapter we focus on $S = 1/2$.

4.3 Low-energy expansion and sum rules

As we have just seen, the forward Compton scattering off a spin-1/2 target is described by two scalar amplitudes, $f(\nu)$ and $g(\nu)$. As first shown by Gell-Mann, Goldberger, and Thirring [1], these amplitudes satisfy the dispersive representations (4.8) and (4.12), respectively.

Consider now their low-energy expansion, i.e. their Taylor expansion in ν. The first term in the expansion is fixed by the classic low-energy theorem [2, 3], which states that the Compton scattering of a 'soft photon' is given entirely by the electromagnetic moments of the target—the electric charge and the magnetic dipole moment[3]. The subsequent terms in the low-energy expansion can be parametrized by various polarizabilities (induced electromagnetic moments) of the target, such that the expansion to next-to-leading order reads as follows:

$$f(\nu) = -\frac{Z^2\alpha}{M} + (\alpha_{\rm E1} + \beta_{\rm M1})\nu^2 + O(\nu^4), \tag{4.14a}$$

$$g(\nu) = -\frac{\alpha\varkappa^2}{2M^2}\nu + \gamma_0\nu^3 + O(\nu^5), \tag{4.14b}$$

where Z is the charge in units of the proton charge $e = \sqrt{4\pi\alpha}$, and $\varkappa$ is the anomalous magnetic moment of the target. Furthermore, $\alpha_{\rm E1}$ and $\beta_{\rm M1}$ are, respectively, the electric and magnetic dipole polarizabilities, and γ_0 is the so-called forward-spin polarizability (FSP)[4].

These expansions can be substituted on the left-hand side of equations (4.8) and (4.12), whereas the right-hand side can be expanded using $1/(\nu'^2 - \nu^2) = 1/\nu'^2 + \nu^2/\nu'^4 + O(\nu^4)$. Matching the terms of the same order in ν on the left and the right we obtain the sum rules. The caveat, much discussed in chapter 2, is that the first term in the expansion of $f(\nu)$ yields nonsense:

$$-\frac{Z^2\alpha}{M} = \frac{2}{\pi}\int_0^\infty d\nu\, \sigma(\nu). \tag{4.15}$$

This can never work, as the left-hand side is negative whereas the right-hand side is positive. As discussed in chapter 2, this failure is caused by a violation of the convergence properties which go into the derivation of the dispersion relation (namely, the non-vanishing $f(\nu)$ for asymptotically large energy). The problem is circumvented by applying a subtraction at $\nu = 0$ to obtain

$$\boxed{f(\nu) = -\frac{Z^2\alpha}{M} + \frac{\nu^2}{2\pi^2}\int_0^\infty d\nu'\, \frac{\sigma(\nu')}{\nu'^2 - \nu^2 - i0^+}}. \tag{4.16}$$

[3] A particle with spin S has in general $2S + 1$ electromagnetic moments.

[4] The FSP is the following linear combination of the four spin polarizabilities appearing at this order (see e.g. [4–6] for more details): $\gamma_0 = -(\gamma_{\rm E1E1} + \gamma_{\rm M1M1} + \gamma_{\rm E1M2} + \gamma_{\rm M1E2})$.

The price to pay for having the subtraction is the loss of a potential sum rule for the electric charge. Well, it looks like the nature prevents us from calculating the electric charge. We are only allowed to measure it.

Fortunately, all the other sum rules are working out famously. At $O(\nu^2)$ in the low-energy expansion of $f(\nu)$, one obtains the Baldin sum rule [7]

$$\boxed{\alpha_{\mathrm{E1}} + \beta_{\mathrm{M1}} = \frac{1}{2\pi^2}\int_0^\infty d\nu\,\frac{\sigma(\nu)}{\nu^2}.} \tag{4.17}$$

At $O(\nu)$ in the low-energy expansion of $g(\nu)$, one obtains the celebrated Gerasimov–Drell–Hearn (GDH) sum rule [8–10]:

$$\boxed{\frac{\alpha}{M^2}\varkappa^2 = -\frac{1}{2\pi^2}\int_0^\infty d\nu\,\frac{\sigma_{1/2}(\nu) - \sigma_{3/2}(\nu)}{\nu}.} \tag{4.18}$$

This sum rule, expressing the anomalous magnetic moment (squared!) in terms of an energy-weighted integral of the helicity-difference photoabsorption cross section, is possibly the best known example of sum rules. It directly demonstrates the idea of expressing purely quantum effects, which give rise to the anomalous magnetic moment, in terms of an integral of a classical quantity—the cross section. This is somewhat analogous to the idea of path-integral quantization in ordinary quantum mechanics (or, quantum field theory), where the quantum transition amplitude arises as an integral over all classical trajectories (field configurations).

The difference here is that the cross section is an experimentally observable quantity. If we manage to measure it, channel by channel, we obtain a lot of useful information about the composition of effects that generate the anomalous magnetic moment. Exactly the same statement can be made about the Schwinger sum rule, considered in the next chapter (section 5.4), except that it is even better suited for this task, because there the anomalous magnetic moment enters linearly. The perturbative verifications of the GDH sum rule in QED and other quantum field theories are also more involved than those for the Schwinger sum rule. Nevertheless they do provide new insights into the nature of this sum rule, see e.g. [11–14].

At $O(\nu^3)$ in the low-energy expansion of $g(\nu)$, one arrives at the sum rule for the forward-spin polarizability, sometimes also referred to as a Gell-Mann, Goldberger, and Thirring (GGT) sum rule:

$$\boxed{\gamma_0 = \frac{1}{4\pi^2}\int_0^\infty d\nu\,\frac{\sigma_{1/2}(\nu) - \sigma_{3/2}(\nu)}{\nu^3}.} \tag{4.19}$$

As this would not be the only GTT sum rule (see e.g. [15]), we shall refer to equation (4.19) as the FSP sum rule.

The low-energy expansion of equation (4.14) can of course be continued, leading to as many sum rules as one wants. The only problem is to express the coefficients in terms of some useful physical quantities. At $O(\nu^4)$, such quantities are the energy slopes of the dipole polarizabilities ($\alpha_{\mathrm{E1}\nu}$, $\beta_{\mathrm{M1}\nu}$) and the quadrupole polarizabilities (α_{E2}, β_{M2}):

$$\boxed{\alpha_{E1\nu} + \beta_{M1\nu} + \frac{1}{12}(\alpha_{E2} + \beta_{M2}) = \frac{1}{2\pi^2}\int_0^\infty d\nu\, \frac{\sigma(\nu)}{\nu^4}.} \tag{4.20}$$

We refer to it as 'the fourth-order Baldin sum rule'.

4.4 Empirical evaluations for the nucleon

The above sum rules are well-studied for the nucleon, i.e. the proton ($Z = 1$) and the neutron ($Z = 0$). Let's consider this case as an example of how all of it works in practice and why the sum rules are useful.

First of all, one needs the two photoabsorption cross sections, the unpolarized and the helicity-difference cross sections:

$$\sigma(\nu) = \frac{1}{2}[\sigma_{3/2}(\nu) + \sigma_{1/2}(\nu)], \tag{4.21a}$$

$$\Delta\sigma(\nu) = \sigma_{3/2}(\nu) - \sigma_{1/2}(\nu). \tag{4.21b}$$

Both of them are fairly well known for the proton, with a lot of new precise data coming in the past two decades from the Mainz and Bonn facilities MAMI and ELSA (see e.g. [29–31]). The current situation, according Gryniuk *et al* [27, 28], is shown in figure 4.2, both unpolarized (upper panel) and polarized (lower panel).

The idea, which goes back to Damashek and Gilman [32], is to fit the available experimental data with a physically motivated parametrization, obtaining an integrable function which can then be used to evaluate the above sum rules, see tables 4.1 and 4.2. The tables also contain some results for the neutron.

One can see that the sum rules provide a precise determination of the forward combinations of nucleon polarizabilities. Currently the Baldin sum sum rule determination of the sum of the electric and magnetic polarizabilities for the proton is about an order of magnitude more precise than direct measurements of these quantities in Compton scattering experiments (see [6, 41, 42] for recent reviews).

As for the GDH sum rule evaluation in table 4.2, the situation is different. The most precise determination of the nucleon anomalous magnetic moment is of course coming from the direct measurement. Feeding in the experimental value of the proton and neutron anomalous magnetic moments ($\varkappa_p \simeq 1.793$, $\varkappa_n \simeq -1.913$) into the left-hand side (lhs) of the GDH sum rule (4.18), we obtain the first row of table 4.2. The evaluation of the right-hand side (rhs) using the empirical helicity-difference cross section is many orders of magnitude less precise. Nevertheless, they are at least consistent with each other. The most recent evaluation [28] shows that the GDH sum rule for the proton is verified to 2% accuracy.

The empirical parametrizations of the photoabsorption cross sections can also be used to determine the entire Compton scattering amplitude. For example, the unpolarized amplitude for the proton is shown in figure 4.3, where it can be compared with the results of previous evaluations and an experimental point from the DESY 1973 experiment [37], see also the last column of table 4.1. In the latter experiment the forward Compton scattering amplitude was accessed indirectly via

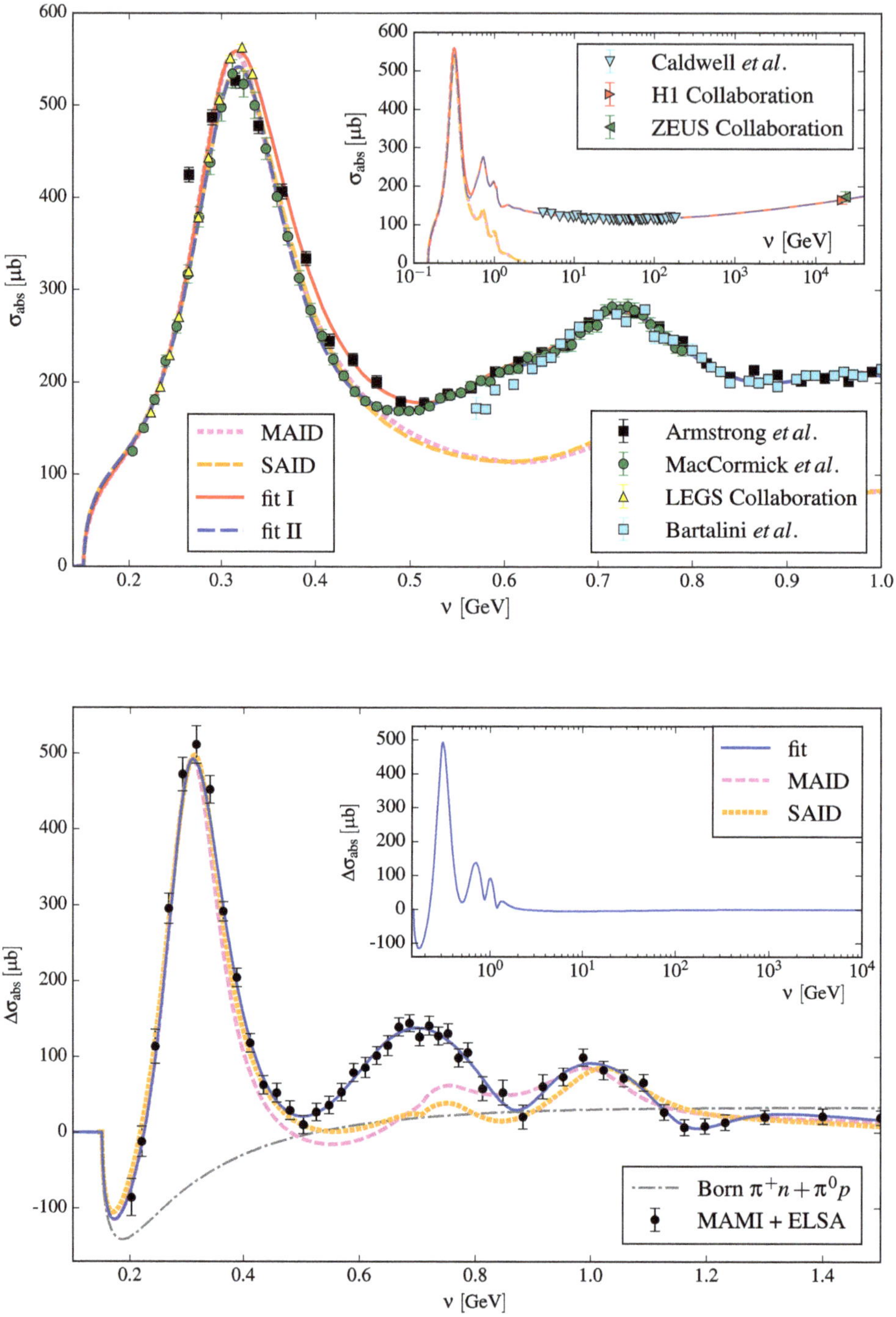

Figure 4.2. Total photoabsorption on the proton. *Upper panel:* unpolarized cross section. Fit I includes MAID [16] results below the 2π production, and data from LEGS [17] and Armstrong *et al* [18] above the threshold; fit II includes SAID [19] and the data of MacCormick *et al* [20]. Both fits use Bartalini *et al* [21] and the high-energy data [22–24] displayed in the inset. *Lower panel:* helicity-difference cross section. The solid curve is the parametrization of experimental data. The experimental data are from the GDH Collaboration [25, 26]. For further details, see [27, 28].

Table 4.1. Empirical evaluation of spin-independent sum rules for the proton and neutron.

	Baldin $[10^{-4}\ \mathrm{fm}^3]$		Fourth order $[10^{-4}\ \mathrm{fm}^5]$		Sixth order $[10^{-4}\ \mathrm{fm}^7]$	Re f at 2.2 GeV $[\mu\mathrm{b\ GeV}]$
Reference	Proton	Neutron	Proton	Neutron	Proton	Proton
Damashek *et al* [32]	14.2 ± 0.3					
Schroder [33]	14.7 ± 0.7	13.3 ± 0.7	6.4	5.6		
Levchuk *et al* [34]	14.0(3…5)	15.2 ± 0.4				
Babusci *et al* [35]	13.69 ± 0.14	14.40 ± 0.66				
A2 Coll. [36]	13.8 ± 0.4					−10.5
DESY exp. [37]						−12.3 ± 2.4
Gryniuk *et al* [27]	14.00 ± 0.20		6.04 ± 0.03		4.39 ± 0.03	−10.18

Table 4.2. Empirical evaluation of the GDH and FSP sum rules for the proton and neutron.

	GDH integral $[\mu\mathrm{b}]$		$\gamma_0\ [10^{-4}\ \mathrm{fm}^4]$	
Reference	Proton	Neutron	Proton	Neutron
Sum rule lhs	205	233		
GDH Coll. [26, 30, 31]	212 ± 17	225	−1.01 ± 0.13	
Gryniuk *et al* [28]	204.5 ± 9.4		−0.93 ± 0.06	
Pasquini *et al* [38]	210 ± 15		−0.90 ± 0.14	
Babusci *et al* [39]			−1.5	−0.4
Schumacher *et al* [40]			−0.58 ± 0.20	0.38 ± 0.22

the process of dilepton photoproduction. A direct measurement of the forward Compton scattering amplitude is not possible, at least using the present technology. On the other hand, as we can see, the dispersion relation becomes an instrument for an accurate empirical determination of the forward amplitude. This is precisely why the dispersive methods are useful.

Having obtained in this way the amplitudes $f(\nu)$ and $g(\nu)$, one can reconstruct the experimental observables for the forward Compton scattering. There are two of them—the unpolarized differential cross section and the double-spin asymmetry:

$$\frac{d\sigma}{d\Omega_{\mathrm{lab}}} \overset{t=0}{=} |f|^2 + |g|^2, \qquad \Sigma_{2z} \overset{t=0}{=} -\frac{fg^* + f^*g}{|f|^2 + |g|^2}. \tag{4.22}$$

All the information about the proton Compton scattering is thus fully determined by the photoabsorption cross sections.

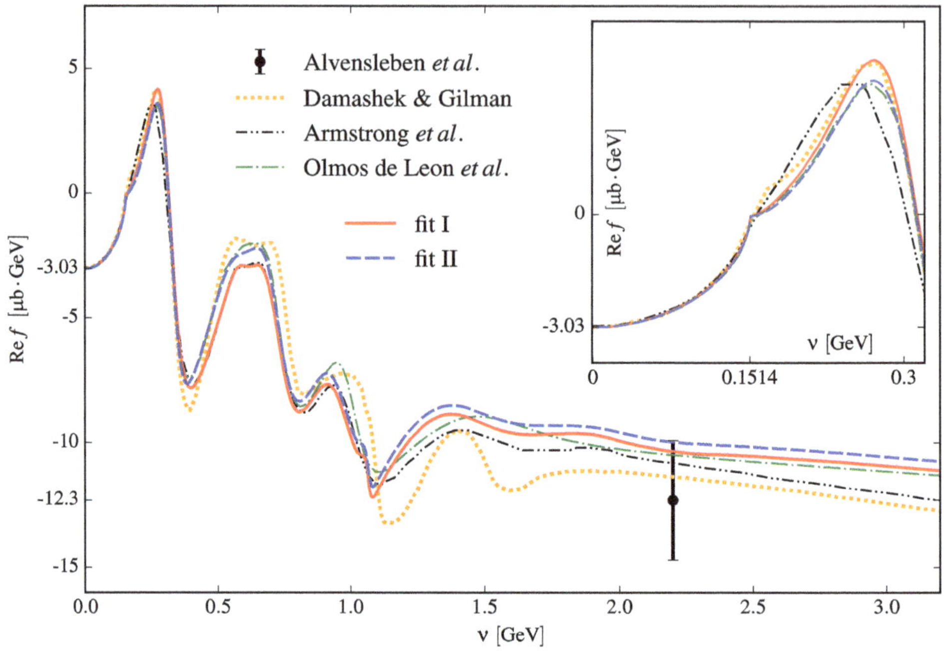

Figure 4.3. Amplitude $f(\nu)$ for the proton obtained from equation (4.16) using different fits of the total photoabsorption cross section [18, 27, 32, 36] shown in figure 4.2. Fits I and II refer to the results of reference [27]. The data point at 2.2 GeV is from [37].

References

[1] Gell-Mann M, Goldberger M and Thirring W E 1954 Use of causality conditions in quantum theory *Phys. Rev.* **95** 1612–27

[2] Low F E 1954 Scattering of light of very low frequency by systems of spin 1/2 *Phys. Rev.* **96** 1428–32

[3] Gell-Mann M and Goldberger M L 1954 Scattering of low-energy photons by particles of spin 1/2 *Phys. Rev.* **96** 1433–8

[4] Guiasu I and Radescu E 1979 Higher multipole polarizabilities of hadrons from Compton scattering amplitudes *Annals Phys.* **120** 145

[5] Drechsel D, Pasquini B and Vanderhaeghen M 2003 Dispersion relations in real and virtual Compton scattering *Phys. Rep.* **378** 99–205

[6] Hagelstein F, Miskimen R and Pascalutsa V 2016 Nucleon polarizabilities: from Compton scattering to hydrogen atom *Prog. Part. Nucl. Phys.* **88** 29–97

[7] Baldin A M 1960 Polarizability of nucleons *Nucl. Phys.* **18** 310–7

[8] Gerasimov S 1966 A sum rule for magnetic moments and the damping of the nucleon magnetic moment in nuclei *Sov. J. Nucl. Phys.* **2** 430–3

[9] Drell S and Hearn A C 1966 Exact sum rule for nucleon magnetic moments *Phys. Rev. Lett.* **16** 908–11

[10] Hosoda M and Yamamoto K 1966 Sum rule for the magnetic moment of the Dirac particle *Prog. Theor. Phys.* **36** 425–6

[11] Altarelli G, Cabibbo N and Maiani L 1972 The Drell–Hearn sum rule and the lepton magnetic moment in the Weinberg model of weak and electromagnetic interactions *Phys. Lett.* B **40** 415

[12] Dicus D A and Vega R 2001 The Drell–Hearn sum rule at order α^3 *Phys. Lett.* B **501** 44–7

[13] Pascalutsa V, Holstein B R and Vanderhaeghen M 2004 A derivative of the Gerasimov–Drell–Hearn sum rule *Phys. Lett.* B **600** 239–47

[14] Holstein B R, Pascalutsa V and Vanderhaeghen M 2005 Sum rules for magnetic moments and polarizabilities in QED and chiral effective-field theory *Phys. Rev.* D **72** 094014

[15] Weise W 1973 Generalized dispersive photonuclear sum rule *Phys. Rev. Lett.* **31** 773–6

[16] Drechsel D, Kamalov S and Tiator L 2007 Unitary Isobar Model—MAID2007 *Eur. Phys. J.* A**34** 69–97 (MAID program: http://www.kph.uni–mainz.de/MAID/)

[17] Blanpied G *et al* 1996 $N \to \Delta$ transition from simultaneous measurements of $p(\overrightarrow{\gamma}, \pi)$ and $p(\overrightarrow{\gamma}, \gamma)$ *Phys. Rev. Lett.* **79** 4337–40

[18] Armstrong T *et al* 1972 Total hadronic cross-section of gamma rays in hydrogen in the energy range 0.265 GeV to 4.215 GeV *Phys. Rev.* D **5** 1640–52

[19] Workman R L, Paris M W, Briscoe W J and Strakovsky I I 2012 Unified Chew–Mandelstam SAID analysis of pion photoproduction data *Phys. Rev.* C **86** 015202 (SAID program: http://gwdac.phys.gwu.edu)

[20] MacCormick M *et al* 1996 Total photoabsorption cross-sections for 1H, 2H and 3He from 200 MeV to 800 MeV *Phys. Rev.* C **53** 41–9

[21] Bartalini O *et al* 2008 Measurement of the total photoabsorption cross section on a proton in the energy range 600–1500 MeV at the GRAAL *Phys. Atom. Nucl.* **71** 75–82

[22] Caldwell D O *et al* 1978 Measurements of the photon total cross-section on protons from 18 GeV to 185 GeV *Phys. Rev. Lett.* **40** 1222

[23] Aid S *et al* 1995 Measurement of the total photon–proton cross-section and its decomposition at 200 GeV center-of-mass energy *Z. Phys.* C **69** 27–38

[24] Chekanov S *et al* 2002 Measurement of the photon proton total cross-section at a center-of-mass energy of 209-GeV at HERA *Nucl. Phys.* B **627** 3–28

[25] Dutz H *et al* 2004 Experimental check of the Gerasimov–Drell–Hearn sum rule for ^{1}H *Phys. Rev. Lett.* **93** 032003

[26] Helbing K 2006 The Gerasimov–Drell–Hearn sum rule *Prog. Part. Nucl. Phys.* **57** 405–69

[27] Gryniuk O, Hagelstein F and Pascalutsa V 2015 Evaluation of the forward Compton scattering off protons: spin-independent amplitude *Phys. Rev.* D **92** 074031

[28] Gryniuk O, Hagelstein F and Pascalutsa V 2016 Evaluation of the forward Compton scattering off protons: II. Spin-dependent amplitude and observables *Phys. Rev.* D **94** 034043

[29] MacCormick M *et al* 1997 Total photoabsorption cross-section for ^{4}He from 200 MeV to 800 MeV *Phys. Rev.* C **55** 1033–8

[30] Ahrens J *et al* 2001 First measurement of the Gerasimov–Drell–Hearn integral for hydrogen from 200 to 800 MeV *Phys. Rev. Lett.* **87** 022003

[31] Dutz H *et al* 2003 First measurement of the Gerasimov–Drell–Hearn sum rule for ^{1}H from 0.7 GeV to 1.8 GeV at ELSA *Phys. Rev. Lett.* **91** 192001

[32] Damashek M and Gilman F J 1970 Forward Compton scattering *Phys. Rev.* D **1** 1319–32

[33] Schröder U 1980 Calculation of the electric polarizabilities of proton and neutron *Nucl. Phys.* B **166** 103

[34] Levchuk M I and L'vov A I 2000 Deuteron Compton scattering below pion photoproduction threshold *Nucl. Phys.* A **674** 449–92

[35] Babusci D, Giordano G and Matone G 1998 A new evaluation of the Baldin sum rule *Phys. Rev.* C **57** 291–4
[36] Olmos de León V *et al* 2001 Low-energy Compton scattering and the polarizabilities of the proton *Eur. Phys. J.* A **10** 207–15
[37] Alvensleben H *et al* 1973 Experimental verification of the Kramers–Kronig relation at high energy *Phys. Rev. Lett.* **30** 328–32
[38] Pasquini B, Pedroni P and Drechsel D 2010 Higher order forward spin polarizability *Phys. Lett.* B **687** 160–6
[39] Babusci D, Giordano G, L'vov A I, Matone G and Nathan A M 1998 Low-energy Compton scattering of polarized photons on polarized nucleons *Phys. Rev.* C **58** 1013–41
[40] Schumacher M and Levchuk M 2011 Structure of the nucleon and spin-polarizabilities *Nucl. Phys.* A **858** 48–66
[41] Grießhammer H, McGovern J, Phillips D and Feldman G 2012 Using effective field theory to analyse low-energy Compton scattering data from protons and light nuclei *Prog. Part. Nucl. Phys.* **67** 841–97
[42] Holstein B R and Scherer S 2014 Hadron polarizabilities *Ann. Rev. Nucl. Part. Sci.* **64** 51–81

Chapter 5

Virtual Compton scattering and quasi-real sum rules

Some of the most intriguing and useful relations and sum rules are derived from the case when the photon involved in the forward Compton scattering is virtual. This process is referred to as the doubly virtual Compton scattering (VVCS). When the photon virtuality Q^2 is infinitesimally small, the photon is called 'quasi-real'. In this limit, for example, one obtains the Schwinger sum rule for the anomalous magnetic moment, along with several elaborate relations for polarizabilities. In what follows, we sketch the derivation of these sum rules and relations, and discuss their empirical consequences and potential applications.

5.1 VVCS and structure functions

The forward VVCS is a central process in studying deep inelastic scattering (DIS), by means of which the proton substructure—quarks and gluons—was discovered. In DIS, the electron probes a proton, or a heavier nucleus, at a high momentum-transfer $Q^2 = -(k - k')^2$ carried by the exchanged photon, see figure 5.1. In this one-photon exchange picture, it is possible to separate out the electron line and focus on the virtual photon striking the target. The square of such an amplitude is identical to a cut-in-half diagram of VVCS, see figure 5.2. Hence, the absorptive part of the VVCS amplitude is related to the DIS observable in a kind of optical theorem. The rest is easy. Employing the analyticity properties of the VVCS amplitude, one can express the entire amplitude as an integral of the observables and then study its low-energy behavior to arrive at sum rules.

To put this in formulae, let us start with the forward VVCS amplitude for the spin-1/2 case. Its Lorentz structure is split into symmetric and antisymmetric

doi:10.1088/978-0-7503-3431-0ch5

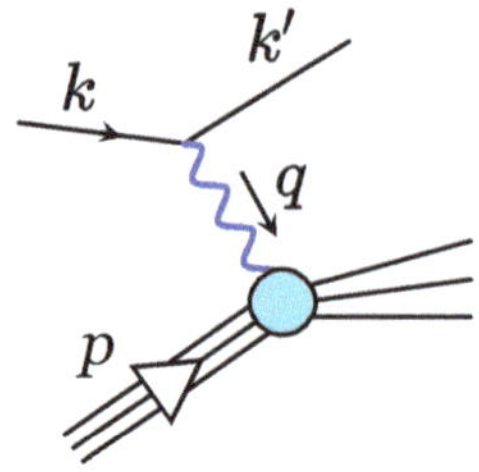

Figure 5.1. Inelastic electron scattering in the one-photon exchange approximation.

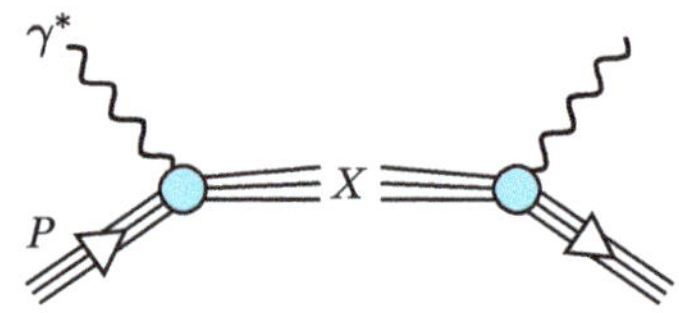

Figure 5.2. Virtual-photoabsorption amplitude squared, where γ^* stands for the virtual photon, P for the target particle (e.g. a proton), and X for any possible state.

(with respect to the vector indices of the initial and final photon) parts as follows[1]:

$$T^{\mu\nu} = T_S^{\mu\nu} + T_A^{\mu\nu},$$

$$T_S^{\mu\nu} = \left(-g^{\mu\nu} + \frac{q^\mu q^\nu}{q^2}\right) T_1(\nu, Q^2) + \left(p^\mu - \frac{p\cdot q}{q^2}\, q^\mu\right)\left(p^\nu - \frac{p\cdot q}{q^2}\, q^\nu\right)\frac{T_2(\nu, Q^2)}{M^2}, \quad (5.1a)$$

$$T_A^{\mu\nu} = \frac{i}{M}\,\epsilon^{\mu\nu\alpha\beta}\, q_\alpha s_\beta\, S_1(\nu, Q^2) + \frac{i}{M^3}\,\epsilon^{\mu\nu\alpha\beta}\, q_\alpha(p\cdot q\; s_\beta - s\cdot q\; p_\beta)\, S_2(\nu, Q^2), \quad (5.1b)$$

where $g^{\mu\nu} = \text{diag}(1, -1, -1, -1)$ is the metric tensor, $\varepsilon_{0123} = +1$, and s^α is the spin four-vector which describes the target polarization, satisfying $s^2 = -1$ and $s\cdot p = 0$. The antisymmetric part is thus spin-dependent, whereas the symmetric part is not. This form is explicitly gauge invariant, since

$$q_\mu T^{\mu\nu} = 0 = q_\nu T^{\mu\nu}, \quad (5.2)$$

which implies the electromagnetic current conservation. The scalar amplitudes $T_{1,\,2}$ (spin-independent) and $S_{1,\,2}$ (spin-dependent) are functions of the photon invariant

[1] So far, we have managed to avoid the introduction of Dirac matrices and spinors in the discussion of the spin-1/2 case. For those who are familiar with these concepts, there is a useful form for the spin-dependent part of the VVCS amplitude:

$$T_A^{\mu\nu} = \bar{u}(p)\left[-\frac{1}{M}\gamma^{\mu\nu\alpha}q_\alpha\, S_1(\nu, Q^2) + \frac{Q^2}{M^2}\gamma^{\mu\nu}S_2(\nu, Q^2)\right]u(p),$$

with $\gamma^{\mu\nu} = \gamma^\mu\gamma^\nu - g^{\mu\nu}$, $\gamma^{\mu\nu\alpha} = \frac{1}{2}(\gamma^\mu\gamma^\nu\gamma^\alpha - \gamma^\alpha\gamma^\nu\gamma^\mu)$ being the fully antisymmetrized products of Dirac matrices.

energy $\nu = p \cdot q/M = (s - u)/4M$ and its space–like virtuality $Q^2 = -q^2$. Furthermore, $t = 0$ in the forward kinematics, and hence $s + u = 2M^2 - 2Q^2$.

One immediate observation, following from (5.1a), is that the symmetry under photon crossing translates into the following conditions on the scalar amplitudes:

$$T_1(-\nu, Q^2) = T_1(\nu, Q^2), \quad T_2(-\nu, Q^2) = T_2(\nu, Q^2), \tag{5.3a}$$

$$S_1(-\nu, Q^2) = S_1(\nu, Q^2), \quad S_2(-\nu, Q^2) = -S_2(\nu, Q^2). \tag{5.3b}$$

Hence, S_2 is odd with respect to the sign reflection of ν, while the other amplitudes are even. One often considers the combination νS_2, such that it has the same crossing properties as the other amplitudes.

The above-mentioned analogue of the optical theorem relates the absorptive parts of the forward VVCS amplitudes to the *structure functions*, or equivalently, to the cross sections of virtual-photon absorption $\gamma^* N \to X$:

$$\text{Im } T_1(\nu, Q^2) = \frac{4\pi^2\alpha}{M} F_1(x, Q^2) = \nu\, \sigma_T(\nu, Q^2), \tag{5.4a}$$

$$\text{Im } T_2(\nu, Q^2) = \frac{4\pi^2\alpha}{\nu} F_2(x, Q^2) = \frac{Q^2\nu}{\nu^2 + Q^2}[\sigma_T + \sigma_L](\nu, Q^2), \tag{5.4b}$$

$$\text{Im } S_1(\nu, Q^2) = \frac{4\pi^2\alpha}{\nu}\, g_1(x, Q^2) = \frac{M\nu^2}{\nu^2 + Q^2}\left[\frac{Q}{\nu}\sigma_{LT} + \sigma_{TT}\right](\nu, Q^2), \tag{5.4c}$$

$$\text{Im } S_2(\nu, Q^2) = \frac{4\pi^2\alpha M}{\nu^2}\, g_2(x, Q^2) = \frac{M^2\nu}{\nu^2 + Q^2}\left[\frac{\nu}{Q}\sigma_{LT} - \sigma_{TT}\right](\nu, Q^2). \tag{5.4d}$$

Note that the structure functions $F_{1,2}$ (unpolarized) and $g_{1,2}$ (polarized) are customarily written as functions of the Bjorken variable,

$$x = \frac{Q^2}{2M\nu}, \quad \text{with } x \in [0, 1], \tag{5.5}$$

rather than the energy ν. The condition of x being within the unit interval corresponds to the physical domain, $s \geqslant M^2$, hence $2M\nu \geqslant Q^2$. The lowest value of the energy here is $\nu_{el} = Q^2/2M$, referred to as the 'elastic threshold'. In the DIS limit ($Q^2 \to \infty$, fixed x), the structure functions are related to the *parton distribution functions*. We, however, are mainly interested in the limit of low Q, where, as we shall see, the moments of structure functions represent static electromagnetic properties, such as polarizabilities.

The cross sections, more properly referred to as 'response functions' because the photons are off-shell (finite Q^2), are defined in terms of the helicity cross sections as $\sigma_T = 1/2\,(\sigma_{1/2} + \sigma_{3/2})$ and $\sigma_{TT} = 1/2\,(\sigma_{1/2} - \sigma_{3/2})$ for transversely polarized photons, $\sigma_L = 1/2\,(\sigma_{1/2} + \sigma_{-1/2})$ for longitudinally polarized photons, and σ_{LT} for the situation when the photon helicity flips from longitudinal to transverse, accompanied by a spin-flip of the target, such that the total helicity is conserved. The response functions σ_T and σ_{TT} in the real-photon limit ($Q^2 = 0$) give, respectively, the unpolarized and the helicity-difference cross sections discussed in the previous chapter. The other two functions vanish in the real-photon limit as $\sigma_L \sim Q^2$ and $\sigma_{LT} \sim Q$.

We next assume the standard analytic structure of the VVCS amplitudes $T_{1,2}$, S_1, and νS_2 in the energy ν, which implies the familiar dispersion relations for the even functions:

$$\{T_1, T_2, S_1, \nu S_2\}(\nu, Q^2) = \frac{2}{\pi} \int_{\nu_{el}}^{\infty} d\nu' \, \nu' \, \frac{\mathrm{Im}\{T_1, T_2, S_1, \nu' S_2\}(\nu', Q^2)}{\nu'^2 - \nu^2 - i0^+}. \tag{5.6a}$$

Substituting in here the unitarity relations (5.4), we obtain[2]

$$\begin{aligned} T_1(\nu, Q^2) &= \frac{8\pi\alpha}{M} \int_0^1 \frac{dx}{x} \frac{F_1(x, Q^2)}{1 - x^2(\nu/\nu_{el})^2 - i0^+} \\ &= \frac{2}{\pi} \int_{\nu_{el}}^{\infty} d\nu' \, \frac{\nu'^2 \sigma_T(\nu', Q^2)}{\nu'^2 - \nu^2 - i0^+}, \end{aligned} \tag{5.8a}$$

$$\begin{aligned} T_2(\nu, Q^2) &= \frac{16\pi\alpha M}{Q^2} \int_0^1 dx \, \frac{F_2(x, Q^2)}{1 - x^2(\nu/\nu_{el})^2 - i0^+} \\ &= \frac{2Q^2}{\pi} \int_{\nu_{el}}^{\infty} d\nu' \, \frac{\nu'^2 [\sigma_T + \sigma_L](\nu', Q^2)}{(\nu'^2 + Q^2)(\nu'^2 - \nu^2 - i0^+)}, \end{aligned} \tag{5.8b}$$

$$\begin{aligned} S_1(\nu, Q^2) &= \frac{16\pi\alpha M}{Q^2} \int_0^1 dx \, \frac{g_1(x, Q^2)}{1 - x^2(\nu/\nu_{el})^2 - i0^+} \\ &= \frac{2M}{\pi} \int_{\nu_{el}}^{\infty} d\nu' \, \frac{\nu'^3 \left[\frac{Q}{\nu'} \sigma_{LT} + \sigma_{TT}\right](\nu', Q^2)}{(\nu'^2 + Q^2)(\nu'^2 - \nu^2 - i0^+)}, \end{aligned} \tag{5.8c}$$

$$\begin{aligned} \nu S_2(\nu, Q^2) &= \frac{16\pi\alpha M^2}{Q^2} \int_0^1 dx \, \frac{g_2(x, Q^2)}{1 - x^2(\nu/\nu_{el})^2 - i0^+} \\ &= \frac{2M^2}{\pi} \int_{\nu_{el}}^{\infty} d\nu' \, \frac{\nu'^3 \left[\frac{\nu'}{Q} \sigma_{LT} - \sigma_{TT}\right](\nu', Q^2)}{(\nu'^2 + Q^2)(\nu'^2 - \nu^2 - i0^+)}. \end{aligned} \tag{5.8d}$$

[2] With $x = \nu_{el}/\nu$, $\nu_{el} = Q^2/2M$, the change of the integration variable from ν to x goes as

$$\int_{\nu_{el}}^{\infty} \frac{d\nu}{\nu^n} f(\nu, Q^2) = \left(\frac{2M}{Q^2}\right)^{n-1} \int_0^1 dx \, x^{n-2} f(x, Q^2). \tag{5.7}$$

We are now in a position to consider the various sum rules arising from the low-energy (and, sometimes, low-Q) expansion of these amplitudes. Note that, in the real-photon limit, we simply recover the results of the previous chapter by identifying the two real-Compton scattering amplitudes as

$$f(\nu) = \frac{T_1(\nu, 0)}{4\pi}, \qquad g(\nu) = \frac{\nu}{M}\frac{S_1(\nu, 0)}{4\pi}. \tag{5.9}$$

In what follows, we will see what can be gained by keeping Q finite.

5.2 Elastic versus Born contributions

There is one more hurdle on the way to the sum rules—the low-energy theorem for the VVCS amplitudes. As in any derivation of the low-energy theorem, one begins with separating the amplitudes into the Born contribution, where the intermediate state is the particle itself, see figure 5.3, and the rest, called the non-Born piece.

The vertices in the diagrams of figure 5.3 are parametrized in terms of the elastic form factors, which are functions of Q^2. In the case of spin-1/2 particles, there are two form factors (in general, $2S + 1$): the Dirac $F_1(Q^2)$ and Pauli $F_2(Q^2)$ form factors (not to be confused with the unpolarized structure functions!), or their frequently used linear combinations called the electric and magnetic Sachs form factors,

$$G_E(Q^2) = F_1(Q^2) - \tau F_2(Q^2), \quad G_M(Q^2) = F_1(Q^2) + F_2(Q^2), \quad \text{with } \tau = \frac{Q^2}{4M^2} \tag{5.10}$$

At $Q^2 = 0$ they are normalized to the charge and the anomalous magnetic moment as

$$F_1(0) = G_E(0) = Z, \quad F_2(0) = G_M(0) - Z = \varkappa. \tag{5.11}$$

In what follows, we take $Z = 1$ for simplicity.

With these definitions, the Born contribution to the VVCS amplitudes is given by

$$T_1^{\rm Born}(\nu, Q^2) = \frac{4\pi\alpha}{M}\left[\frac{\nu_{\rm el}^2}{\nu_{\rm el}^2 - \nu^2}\, G_M^2(Q^2) - F_1^2(Q^2)\right], \tag{5.12a}$$

$$T_2^{\rm Born}(\nu, Q^2) = \frac{8\pi\alpha\nu_{\rm el}}{\nu_{\rm el}^2 - \nu^2}[F_1^2(Q^2) + \tau\, F_2^2(Q^2)], \tag{5.12b}$$

$$S_1^{\rm Born}(\nu, Q^2) = \frac{2\pi\alpha}{M}\left[\frac{2M\nu_{\rm el}}{\nu_{\rm el}^2 - \nu^2}\, F_1(Q^2)\, G_M(Q^2) - F_2^2(Q^2)\right], \tag{5.12c}$$

Figure 5.3. The Born graphs.

$$S_2^{\mathrm{Born}}(\nu, Q^2) = -\frac{2\pi\alpha\nu}{\nu_{\mathrm{el}}^2 - \nu^2} F_2(Q^2)\, G_{\mathrm{M}}(Q^2). \tag{5.12d}$$

The low-energy interpretation of this result seems to be ambiguous. The soft-photon limit implies both ν and Q^2 are small, but the order in which the limits $\nu \to 0$ and $Q^2 \to 0$ are taken in (5.12a) matters — the limits do not commute for some of the amplitudes (namely, T_1 and S_1). Given that the physical process occurs for $s \geqslant M^2$, or equivalently, $\nu \geqslant Q^2/2M$, the order $Q^2 \to 0$, then $\nu \to 0$ is preferred. Moreover, only in this order does one reproduce the classic soft-photon theorem for Compton scattering [1, 2].

There is another issue which looks problematic. We see that all of the amplitudes exhibit a pole at $\nu = \nu_{\mathrm{el}}$, which could be a problem for our dispersion relations: they do not take this pole into account. One could go back and rewrite the dispersion relations for the case with the pole, but there is a simpler workaround.

For virtual photons, the photoabsorption may either be *elastic* or *inelastic*, as shown in figure 5.4. An explicit calculation shows that the structure functions of the purely elastic scattering (figure 5.4(a)) are expressed in terms of the elastic form factors as

$$F_1^{\mathrm{el}}(x, Q^2) = \frac{1}{2} G_{\mathrm{M}}^2(Q^2)\, \delta(1 - x), \tag{5.13a}$$

$$F_2^{\mathrm{el}}(x, Q^2) = \frac{1}{1 + \tau} [G_{\mathrm{E}}^2(Q^2) + \tau G_{\mathrm{M}}^2(Q^2)]\delta(1 - x), \tag{5.13b}$$

$$g_1^{\mathrm{el}}(x, Q^2) = \frac{1}{2} F_1(Q^2)\, G_{\mathrm{M}}(Q^2)\, \delta(1 - x), \tag{5.13c}$$

$$g_2^{\mathrm{el}}(x, Q^2) = -\frac{\tau}{2} F_2(Q^2)\, G_{\mathrm{M}}(Q^2)\, \delta(1 - x), \tag{5.13d}$$

with the Dirac delta-function having the property $\delta(1 - x) = \nu_{\mathrm{el}}\, \delta(\nu - \nu_{\mathrm{el}})$. Substituting the elastic structure functions in the dispersion relations (5.8a), we obtain the following 'pole contributions':

$$T_1^{\mathrm{pole}}(\nu, Q^2) = \frac{4\pi\alpha}{M} \frac{\nu_{\mathrm{el}}^2}{\nu_{\mathrm{el}}^2 - \nu^2} G_{\mathrm{M}}^2(Q^2), \tag{5.14a}$$

Figure 5.4. (a) Elastic and (b) Inelastic parts of the photoabsorption process. In (a) the target particle stays intact (and on its mass shell), whereas (b) describes the production of any other final state.

$$T_2^{\text{pole}}(\nu, Q^2) = 8\pi\alpha \frac{\nu_{\text{el}}}{\nu_{\text{el}}^2 - \nu^2} \frac{G_E^2(Q^2) + \tau G_M^2(Q^2)}{1 + \tau}, \tag{5.14b}$$

$$S_1^{\text{pole}}(\nu, Q^2) = 4\pi\alpha \frac{\nu_{\text{el}}}{\nu_{\text{el}}^2 - \nu^2} F_1(Q^2)\, G_M(Q^2), \tag{5.14c}$$

$$[\nu S_2]^{\text{pole}}(\nu, Q^2) = -2\pi\alpha \frac{\nu_{\text{el}}^2}{\nu_{\text{el}}^2 - \nu^2} F_2(Q^2)\, G_M(Q^2). \tag{5.14d}$$

They correspond nicely with the pole terms of the Born contribution (5.12a). In other words, the dispersion relations work for the pole terms separately. The elastic-scattering channel in the photoabsorption process yields the pole pieces contained in the Born contribution to the VVCS amplitudes. Therefore, the pole terms compensate the elastic-scattering channel and vice versa. One may thus cancel them and deal with only the inelastic photoabsorption (figure 5.4(b)).

The issue of the pole terms in the Born contribution is thus taken care of. One should not forget, however, that the latter has the non-pole pieces too. They have to be added to the non-Born part of the amplitudes, denoted as $\bar{T}_i$ and $\bar{S}_i$, to form the complete non-pole amplitudes ($T^{\text{np}} = T - T^{\text{pole}}$):

$$T_1^{\text{np}}(\nu, Q^2) = -\frac{4\pi\alpha}{M} F_1^2(Q^2) + \bar{T}_1(\nu, Q^2), \tag{5.15a}$$

$$T_2^{\text{np}}(\nu, Q^2) = \bar{T}_2(\nu, Q^2), \tag{5.15b}$$

$$S_1^{\text{np}}(\nu, Q^2) = -\frac{2\pi\alpha}{M} F_2^2(Q^2) + \bar{S}_1(\nu, Q^2), \tag{5.15c}$$

$$\nu S_2^{\text{np}}(\nu, Q^2) = 2\pi\alpha\, F_2(Q^2)\, G_M(Q^2) + \nu \bar{S}_2(\nu, Q^2). \tag{5.15d}$$

These amplitudes obey the dispersion relations with the inelastic structure functions. Note that, for these amplitudes, the order of limits $Q^2 \to 0$ and $\nu \to 0$ is unimportant[3]. We next look at the dispersion relations for the non-pole parts in these various limits.

5.3 The Burkhardt–Cottingham sum rule

Let us start with S_2, because it is the easiest. In the low-energy limit ($\nu \to 0$), the non-Born part of νS_2 vanishes, i.e.

$$\nu \bar{S}_2(\nu, Q^2) \xrightarrow{\nu \to 0} 0. \tag{5.16}$$

[3] The order is only important for the pole terms that vanish in the limit $Q^2 \to 0$, then $\nu \to 0$, as required by soft-photon theorems, but do not vanish in the reverse order.

Otherwise, $\bar{S}_2$ would have a pole at $\nu = 0$, and that is not possible according to the low-energy theorems. Substituting equation (5.15d) into the dispersion relation (5.8d), we therefore have in this limit

$$2\pi\alpha\, F_2(Q^2)\, G_M(Q^2) = \frac{16\pi\alpha M^2}{Q^2} \int_0^{x_0} dx\, g_2(x, Q^2), \tag{5.17}$$

where x_0 is the inelastic threshold, to account for the fact that we consider the non-pole part of the amplitude only. We can of course add the elastic contribution, see equation (5.14d), to both sides of this relation, resulting in the following sum rule:

$$\boxed{0 = \int_0^1 dx\, g_2(x, Q^2).} \tag{5.18}$$

This is the familiar form of the Burkhardt–Cottingham (BC) sum rule [3]. Note that it is valid for any $Q^2 \geqslant 0$. The photon virtuality cannot be time-like, in order to exclude the particle production off the external photons.

Coming back to arbitrary ν, we can see that the non-Born amplitude satisfies a once-subtracted type of dispersion relation:

$$\begin{aligned}\bar{S}_2(\nu, Q^2) &= \frac{64\pi\alpha M^4 \nu}{Q^6} \int_0^{x_0} dx\, \frac{x^2 g_2(x, Q^2)}{1 - x^2(\nu/\nu_{el})^2 - i0^+} \\ &= \frac{2M^2\nu}{\pi} \int_{\nu_0}^{\infty} d\nu'\, \frac{\nu'\left[\frac{\nu'}{Q}\sigma_{LT} - \sigma_{TT}\right](\nu', Q^2)}{(\nu'^2 + Q^2)(\nu'^2 - \nu^2 - i0^+)},\end{aligned} \tag{5.19}$$

with $\nu_0 = Q^2/2x_0$ being the inelastic threshold in photon energy.

5.4 The Schwinger sum rule

Let us now consider the other spin-dependent VVCS amplitude, $S_1(\nu, Q^2)$. The non-Born part of this amplitude vanishes in the soft-photon limit (i.e. ν and $Q^2 \to 0$): $\bar{S}_1(0, 0) = 0$. Therefore, from the dispersion relation (5.8c), written for the non-pole term (5.15c), we have

$$\begin{aligned}-\frac{2\pi\alpha}{M}\varkappa^2 &= \lim_{Q^2 \to 0} \frac{16\pi\alpha M}{Q^2} \int_0^{x_0} dx\, g_1(x, Q^2) \\ &= \frac{2M}{\pi} \int_{\nu_0}^{\infty} d\nu'\, \frac{\sigma_{TT}(\nu', 0)}{\nu'}.\end{aligned} \tag{5.20}$$

This is nothing other than the GDH sum rule, see equation (4.18); so, nothing new. We can still obtain something interesting by essentially combining the GDH and BC sum rules.

For this purpose, consider the following linear combination of the spin amplitudes:

$$
\begin{aligned}
MS_1^{\mathrm{np}}(\nu, Q^2) + \nu S_2^{\mathrm{np}}(\nu, Q^2) &= 2\pi\alpha F_1(Q^2)F_2(Q^2) + M\bar{S}_1(\nu, Q^2) + \nu\bar{S}_2(\nu, Q^2) \\
&= \frac{16\pi\alpha M^2}{Q^2}\int_0^{x_0} dx\, \frac{g_1(x, Q^2) + g_2(x, Q^2)}{1 - x^2(\nu/\nu_{\mathrm{el}})^2 - i0^+} \\
&= \frac{2M^2}{\pi Q}\int_{\nu_0}^{\infty} d\nu'\, \frac{\nu'^2\, \sigma_{\mathrm{LT}}(\nu', Q^2)}{\nu'^2 - \nu^2 - i0^+}.
\end{aligned}
\tag{5.21}
$$

Now, taking the soft-photon limit, we obtain a sum rule carrying the name of Julian Schwinger:

$$
\boxed{
\begin{aligned}
\varkappa &= \lim_{Q^2\to 0}\frac{8M^2}{Q^2}\int_0^{x_0} dx\, [g_1(x, Q^2) + g_2(x, Q^2)] \\
&= \frac{M^2}{\pi^2\alpha}\int_{\nu_0}^{\infty} d\nu' \lim_{Q^2\to 0}\frac{\sigma_{\mathrm{LT}}(\nu', Q^2)}{Q}.
\end{aligned}
}
\tag{5.22}
$$

It is not easy to spot this sum rule in Schwinger's original papers [4, 5], since it is written in terms of the imaginary parts of the amplitudes, and not the observables (structure or response functions). The expression in terms of the response function σ_{LT}, along with the name 'Schwinger's sum rule', first appeared in [6].

Despite being essentially a linear combination of the GDH and BC sum rules, the Schwinger sum rule has an important difference, which makes it more useful in practical applications: the anomalous magnetic moment enters linearly, rather than quadratically. An extended discussion of why this feature is advantageous can be found in [7], where the Schwinger sum rule is proposed for an assessment of the hadronic contributions to $g - 2$ of the muon, see chapter 9. Here we only consider a simple example of how this sum rule works in quantum electrodynamics (QED).

In QED the leading order in an α contribution to the photoabsorption cross section comes from the Compton channel computed at the tree level, as given by the Feynman diagrams in figure 5.5. These diagrams translate into the following expression for the Compton amplitude:

$$
T^{(\mathrm{tree})\mu\nu} = -e^2\left(\gamma^\mu\frac{\not{p} + \not{q} + M}{s - M^2}\gamma^\nu + \gamma^\nu\frac{\not{p}' - \not{q} + M}{u - M^2}\gamma^\mu\right),
\tag{5.23}
$$

where $s = (p + q)^2 = (p' + q')^2$, $u = (p' - q)^2 = (p - q')^2$ are the Mandelstam invariants; p (p'), q (q') denote the four-momenta of the initial (final) fermion and photon, respectively. The fermion mass-shell condition reads: $p^2 = p'^2 = M^2$. The initial photon is virtual, $q^2 = -Q^2$, and the final one is real, $q'^2 = 0$. One then considers the photoabsorption cross sections where the initial fermions and photons have specific polarizations while summing over the polarizations of final particles.

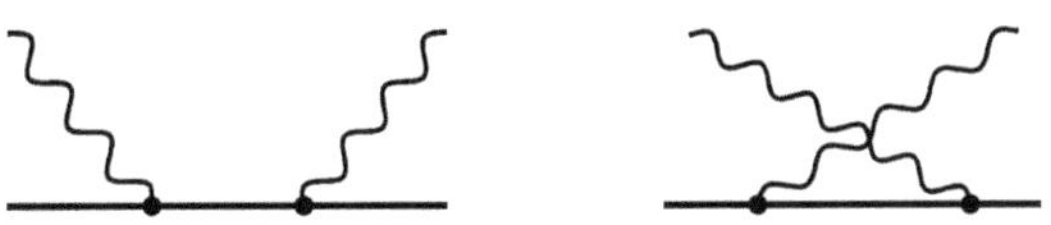

Figure 5.5. Tree-level graphs of Compton scattering in QED.

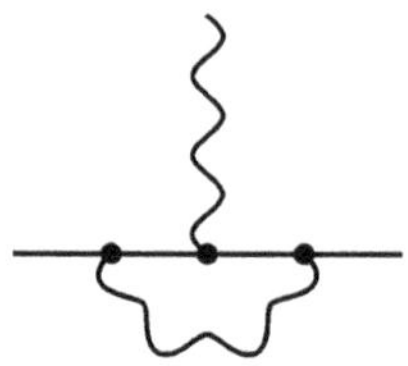

Figure 5.6. Leading-order correction to the electromagnetic vertex, which corresponds to the Schwinger term $\alpha/2\pi$ in the lepton anomalous magnetic moment.

This calculation is not very difficult and gives in the end the following result for the LT response:

$$\sigma_{\rm LT}^{\rm tree}(\nu, Q^2) = \frac{\pi\alpha^2 Q\,(s-M^2)^2}{4M^3\nu^2(\nu^2+Q^2)}\left(-2 - \frac{M(M+\nu)}{s} + \frac{3M+2\nu}{\sqrt{\nu^2+Q^2}}\,\mathrm{arccoth}\frac{M+\nu}{\sqrt{\nu^2+Q^2}}\right) \quad (5.24)$$

with $\nu = p \cdot q/M$. This expression is proportional to Q, which makes it possible to take the quasi-real limit as

$$\left[\frac{\sigma_{\rm LT}^{\rm tree}(\nu, Q^2)}{Q}\right]_{Q^2=0} = \frac{\pi\alpha^2}{M\nu^2}\left(-2 - \frac{M+\nu}{M+2\nu} + \frac{3M+2\nu}{2\nu}\log\frac{M+2\nu}{M}\right) \quad (5.25)$$

Substituting this into the Schwinger sum rule, we obtain

$$\frac{M^2}{\pi^2\alpha}\int_0^\infty d\nu \left[\frac{\sigma_{\rm LT}^{\rm tree}(\nu, Q^2)}{Q}\right]_{Q^2=0} = \frac{\alpha}{2\pi}. \quad (5.26)$$

The right-hand side is the celebrated Schwinger's correction to the anomalous magnetic moment in QED [8], obtained at the one-loop level, see figure 5.6. Thus, the sum rule is verified in QED, at least to leading order. Further details on perturbative QED checks of the various sum rules can be found in [9].

Let us emphasize again that the beauty of the sum rules, such as this, is that they can be used non-perturbatively. So, for example, by measuring the hadronic channels in $\sigma_{\rm LT}$, one could in principle assess their contributions to $\varkappa$, which otherwise require very formidable non-perturbative calculations in QCD. We will discuss this in chapter 9 in more detail.

5.5 Generalized Baldin sum rules

The dispersion relations for the spin-independent amplitudes T_1 and T_2 are no less intriguing than the spin-dependent ones. As already noted in the previous chapter, for $Q^2 = 0$, the convergence properties of the T_1 amplitude are such that its dispersion relations requires a subtraction, see equation (4.16). For finite Q^2, it appears that the dispersion relation for T_1 requires a subtraction at each Q^2. In this case, $T_1(0, Q^2)$ is a 'subtraction function', and the corresponding subtracted dispersion relation, written for the non-pole part, reads as

$$\begin{aligned}T_1^{\rm np}(\nu,\,Q^2) &= T_1^{\rm np}(0,\,Q^2) + \frac{32\pi\alpha M\nu^2}{Q^4}\int_0^{x_0} dx\frac{x\,F_1(x,\,Q^2)}{1 - x^2(\nu/\nu_{\rm el})^2 - i0^+},\\ &= T_1^{\rm np}(0,\,Q^2) + \frac{2\nu^2}{\pi}\int_{\nu_0}^{\infty} d\nu'\,\frac{\sigma_{\rm T}(\nu',\,Q^2)}{\nu'^{\,2} - \nu^2 - i0^+}.\end{aligned} \tag{5.27}$$

For the other amplitude, one may continue using the unsubtracted dispersion relation:

$$\begin{aligned}\bar{T}_2(\nu,\,Q^2) &= \frac{16\pi\alpha M}{Q^2}\int_0^{x_0} dx\,\frac{F_2(x,\,Q^2)}{1 - x^2(\nu/\nu_{\rm el})^2 - i0^+}\\ &= \frac{2Q^2}{\pi}\int_{\nu_0}^{\infty} d\nu'\,\frac{\nu'^{\,2}[\sigma_{\rm T} + \sigma_{\rm L}](\nu',\,Q^2)}{(\nu'^{\,2} + Q^2)(\nu'^{\,2} - \nu^2 - i0^+)},\end{aligned} \tag{5.28}$$

where we have taken into account that the non-pole part is identical to non-Born part for this amplitude, see equation (5.15b).

The combined low-ν and low-Q expansion of the spin-independent amplitudes is given by

$$\begin{aligned}\frac{1}{4\pi}T_1^{\rm np}(\nu,\,Q^2) = &-\frac{\alpha}{M} + \left(\frac{\alpha}{3M}\langle r^2\rangle_1 + \beta_{\rm M1}\right)Q^2 + \left(\alpha_{\rm E1} + \beta_{\rm M1}\right)\nu^2\\ &+ \left[\alpha_{E1\nu} + \beta_{M1\nu} + 1/12\,(\alpha_{E2} + \beta_{M2})\right]\nu^4 + \cdots,\end{aligned} \tag{5.29a}$$

$$\frac{1}{4\pi}\bar{T}_2(\nu,\,Q^2) = \left(\alpha_{\rm E1} + \beta_{\rm M1}\right)Q^2 + \left[\alpha_{E1\nu} + \beta_{M1\nu} + 1/12\,(\alpha_{E2} + \beta_{M2})\right]Q^2\nu^2 + \cdots \tag{5.29b}$$

where, in the case of T_1, the Thomson term and the Dirac radius $\langle r^2\rangle_1 = -6\,d/dQ^2\,F_1(Q^2)|_{Q^2=0}$, come from the Born contribution. The fact that the same combinations of polarizabilities enter in both amplitudes, allows us to write, for instance, the Baldin sum rule in two equivalent ways:

$$\alpha_{\rm E1} + \beta_{\rm M1} = \lim_{Q^2\to 0}\frac{8\alpha M}{Q^4}\int_0^{x_0} dx\,x\,F_1(x,\,Q^2) \tag{5.30a}$$

$$= \lim_{Q^2\to 0}\frac{4\alpha M}{Q^4}\int_0^{x_0} dx\,F_2(x,\,Q^2). \tag{5.30b}$$

At the next order in ν^2, we have the fourth-order sum rule:

$$\alpha_{E1\nu} + \beta_{M1\nu} + 1/12\,(\alpha_{E2} + \beta_{M2}) = \lim_{Q^2\to 0}\frac{32\alpha M^3}{Q^6}\int_0^{x_0} dx\,x^3\,F_1(x,\,Q^2) \tag{5.31a}$$

$$= \lim_{Q^2\to 0}\frac{16\alpha M^3}{Q^6}\int_0^{x_0} dx\,x^2\,F_2(x,\,Q^2). \tag{5.31b}$$

In generalizing these sum rules to finite Q, it makes a difference whether F_1 or F_2 is used. The standard convention is to take the definitions in terms of F_1. For instance, the generalization of the Baldin sum rule,

$$\alpha_{E1}(Q^2) + \beta_{M1}(Q^2) = \frac{8\alpha M}{Q^4} \int_0^{x_0} dx\, x\, F_1(x, Q^2), \tag{5.32}$$

was evaluated for the proton in [10], and more recently in [11, 12], using empirical parametrizations of the structure functions. A more detailed overview of these generalized sum rules can be found in [13, 14], while we now go to a more controversial topic.

5.6 Longitudinal amplitude: to subtract or unsubtract?

Let us consider the Compton scattering of longitudinally polarized photons. The corresponding amplitude is obtained from the general VVCS amplitude (5.1b) by contractions with the longitudinal polarization vectors:

$$T_L(\nu, Q^2) \equiv \varepsilon_L^* \cdot T \cdot \varepsilon_L = -T_1(\nu, Q^2) + \frac{\nu^2 + Q^2}{Q^2} T_2(\nu, Q^2). \tag{5.33}$$

Its imaginary part is, by unitarity, expressed via the longitudinal photoabsorption cross section introduced above,

$$\operatorname{Im} T_L(\nu, Q^2) = \nu \sigma_L(\nu, Q^2). \tag{5.34}$$

The unsubtracted dispersion relation,

$$T_L(\nu, Q^2) = \frac{2}{\pi} \int_{\nu_{el}}^{\infty} d\nu'\, \nu'^2 \frac{\sigma_L(\nu', Q^2)}{\nu'^2 - \nu^2}, \tag{5.35}$$

should, by now, raise no question, other than perhaps the question of convergence. If it does not converge, analogously to what happens with T_1, then it needs a subtraction. If it converges, analogously to T_2, this dispersion relation is good to go.

At first, it looks quite obvious from equation (5.33) that, once T_1 needs a subtraction, T_L needs it too. Remarkably enough, this is not necessarily so. T_1 needs a subtraction when expressed as an integral of σ_T, but what if σ_L is involved as well? Observe that from equation (5.33), taken at $\nu^2 = -Q^2$, one has

$$T_L(iQ, Q^2) = -T_1(iQ, Q^2). \tag{5.36}$$

This is a consequence of Siegert's theorem [15], which equates (up to a phase factor) the transverse and longitudinal amplitudes in the limit of $\nu \to iQ$. Since we know T_1, at the Siegert point, via the unsubtracted dispersion relation (5.35), we can use it as the subtraction point for the T_1 dispersion relation (5.8a) as follows:

$$\begin{aligned} T_1(\nu, Q^2) &= T_1(iQ, Q^2) + \frac{2(\nu^2 + Q^2)}{\pi} \int_{\nu_{el}}^{\infty} d\nu' \frac{\nu'^2}{\nu'^2 + Q^2} \frac{\sigma_T(\nu', Q^2)}{\nu'^2 - \nu^2} \\ &= \frac{2}{\pi} \int_{\nu_{el}}^{\infty} d\nu' \frac{\nu'^2}{\nu'^2 + Q^2} \left[\frac{(\nu^2 + Q^2)\sigma_T(\nu', Q^2)}{\nu'^2 - \nu^2} - \sigma_L(\nu', Q^2) \right]. \end{aligned} \tag{5.37}$$

Thus, we can express the entire amplitude T_1 in terms of both σ_T and σ_L.

Taking here the limit $\nu \to 0$, we obtain a sum rule for the previously unknown subtraction function:

$$T_1(0, Q^2) = \frac{2}{\pi} Q^2 \int_{\nu_{el}}^{\infty} \frac{d\nu}{\nu^2 + Q^2} \left[\sigma_T - \frac{\nu^2}{Q^2} \sigma_L \right] (\nu, Q^2). \tag{5.38}$$

This is very important for the data-driven evaluation of the proton structure effects in muonic hydrogen, to be considered in chapter 10.

On the other hand, the limit of $Q^2 \to 0$ in equation (5.37) and (5.38) requires more care. First, one needs to isolate the elastic contributions. They can be found using equations (5.13a), (5.13b) in conjunction with (5.4), (5.4b), arriving at

$$\sigma_T^{el}(\nu, Q^2) = \frac{2\pi^2 \alpha}{M} G_M^2(Q^2) \delta(\nu - \nu_{el}), \tag{5.39a}$$

$$\sigma_L^{el}(\nu, Q^2) = -\frac{2\pi^2 \alpha}{M} \frac{G_E^2(Q^2)}{\tau} \delta(\nu - \nu_{el}). \tag{5.39b}$$

Recall that $\nu_{el} = Q^2/2M = 2M\tau$. Substituting these elastic cross sections into equation (5.37), we find

$$\begin{aligned} T_1^{el}(\nu, Q^2) &= T_1^{pole}(\nu, Q^2) - \frac{4\pi\alpha}{M} [F_1^2(Q^2) + \tau\, F_2^2(Q^2)] \\ &= T_1^{Born}(\nu, Q^2) - \frac{4\pi\alpha}{M} \tau\, F_2^2(Q^2), \end{aligned} \tag{5.40}$$

where T_1^{pole} is the standard pole term (5.14a). Thus, in this case the elastic contribution generates a non-pole term. This is the main subtlety.

The next step is to replace the lower integration limit ν_{el} by the inelastic threshold ν_0, so as to get rid of the elastic contribution. We only need to ensure that we are doing it on both sides of the relation. On the left-hand side we obtain

$$T_1 - T_1^{el} = T_1^{Born} + \bar{T}_1 - T_1^{el} = \bar{T}_1 + \frac{4\pi\alpha}{M} \tau F_2^2(Q^2), \tag{5.41}$$

where T_1^{Born} is the Born term (5.12a), whereas $\bar{T}_1$ denotes the rest, i.e. the non-Born contribution. Importantly, the Thomson term $T_1(0, 0)$ has disappeared from the equation, and with it, apparently, has disappeared the need for a subtraction. We are left with the following dispersion relation for the non-Born part of T_1:

$$\begin{aligned} \bar{T}_1(\nu, Q^2) =& -\frac{\pi\alpha}{M^3} Q^2 F_2^2(Q^2) \\ &+ \frac{2}{\pi} \int_{\nu_0}^{\infty} d\nu' \frac{\nu'^2}{\nu'^2 + Q^2} \left[\frac{(\nu^2 + Q^2)\sigma_T(\nu', Q^2)}{\nu'^2 - \nu^2} - \sigma_L(\nu', Q^2) \right]. \end{aligned} \tag{5.42}$$

Here the limits of $\nu \to 0$ and $Q^2 \to 0$ are interchangeable and straightforward, since the pole terms are removed.

5.7 The Bernabéu–Tarrach sum rule

Having established the non-Born part of T_1 is fully determined by equation (5.42), we can see what sum rules it generates upon the low-energy expansion. To the leading order in Q^2 and ν^2 we have

$$\bar{T}_1(\nu, Q^2) = 4\pi\left[\beta_{M1}Q^2 + (\alpha_{E1} + \beta_{M1})\nu^2\right] + \cdots \tag{5.43}$$

Equating these with the leading Q^2 and ν^2 terms of the right-hand side of equation (5.42), we obtain, respectively,[4]

$$\beta_{M1} + \frac{\alpha\varkappa^2}{4M^3} = \frac{1}{2\pi^2}\int_0^\infty d\nu\left\{\frac{\sigma_T(\nu)}{\nu^2} - \left[\frac{\sigma_L(\nu, Q^2)}{Q^2}\right]_{Q^2\to 0}\right\}, \tag{5.44}$$

$$\alpha_{E1} + \beta_{M1} = \frac{1}{2\pi^2}\int_0^\infty d\nu\frac{\sigma_T(\nu)}{\nu^2}. \tag{5.45}$$

The first sum rule, involving the magnetic polarizability and the anomalous magnetic moment, appears here for the first time, while the second one should look familiar—it is the Baldin sum rule. Combining them together we can arrive to the following sum rule for the electric dipole polarizability:

$$\boxed{\alpha_{E1} - \frac{\alpha\varkappa^2}{4M^3} = \frac{1}{2\pi^2}\int_0^\infty d\nu\left[\frac{\sigma_L(\nu, Q^2)}{Q^2}\right]_{Q^2\to 0}}. \tag{5.46}$$

It had first been written down in 1975 by Bernabéu and Tarrach [16].

Despite being quite old, the Bernabéu–Tarrach (BT) sum rule has not gained nearly as much popularity as the Baldin sum rule. As seen in the previous chapter, the Baldin sum rule has been instrumental for the data-driven evaluations of the nucleon polarizabilities and provides the most stringent empirical constraint on the sum of proton polarizabilities. In contrast, the BT sum rule was discarded, at least for the nucleons [17] (more recently, in [18]); its use for nuclei was discussed in [19].

Already in 1978, Llanta and Tarrach [20] were the first to discredit the sum rule by showing that it fails a perturbative verification in leading-order QED. They attempted to verify this sum rule in QED by computing the left-hand and right-hand sides (to leading order in α), and found those differ by a constant. This constant, as they show, is given by the value of T_L at $\nu \to \infty$. In chapter 2 we discussed this problem in general terms. The asymptotic contribution at high energy is usually assumed to be vanishing, or divergent (in the latter case the dispersion relation requires subtractions). However, in this perturbative calculation it is apparently finite. Of course, this test by itself does not invalidate the sum rule, since perturbative QED is not exactly valid at arbitrary high energy, but it does cast a shadow over the sum rule applicability.

[4] The smooth $Q^2 \to 0$ limit of σ_L/Q^2 is guaranteed by the electromagnetic gauge invariance.

The available experimental data for σ_L of the proton are quite poor. They are not (yet) sufficiently precise to claim anything about the convergence. Such precise data are highly required for making these BT sum rules as useful empirically as the Baldin sum rule.

It is fairly plausible that the BT sum rule converges, and with it converge the unsubtracted dispersion relations for T_L (5.35) and T_1 (5.42). What works in favor of the BT sum rule, as compared to the failed sum rule for the electric charge (2.20), is that the static quantities on the left-hand side are calculable in theory, whereas the charge is not. Below we consider an elementary example where these sum rules are validated.

5.8 Validation in the parton model

An interesting regime where the relations (5.35), (5.38), and (5.42) could be checked theoretically is the regime of Bjorken scaling. In this limit the virtuality Q^2 and the energy ν of the incoming photon are taken to be very large, while keeping the Bjorken variable $x = Q^2/2M\nu$ finite. The Bjorken scaling implies the validity of the perturbative expansion of QCD, with the leading-order contribution given by the naïve parton model.

In this model the deep inelastic scattering off the proton is described by the scattering off the individual partons (quarks) with electric charges e_q and parton distribution functions $f_q(x)$. The structure functions F_1 and F_2 are then given by (see e.g. section 18.5 in [21])

$$F_1(x, Q^2) = \frac{M}{\alpha}\sum_q e_q^2 f_q(x), \tag{5.47}$$

$$F_2(x, Q^2) = 2xF_1(x, Q^2), \tag{5.48}$$

while the corresponding forward Compton amplitudes are

$$T_1(x, Q^2) = 2\int_0^1 \frac{d\xi}{\xi} \frac{\sum_q e_q^2 f_q(\xi)}{\left(\frac{x}{\xi}\right)^2 - 1 - i\epsilon}, \tag{5.49}$$

$$T_2(x, Q^2) = 8\frac{M^2x^2}{Q^2}\int_0^1 \frac{d\xi}{\xi} \frac{\sum_q e_q^2 f_q(\xi)}{\left(\frac{x}{\xi}\right)^2 - 1 - i\epsilon}. \tag{5.50}$$

The longitudinal structure function F_L is given in this model by

$$\begin{aligned} F_L(x, Q^2) &= \left(1 + \frac{4M^2x^2}{Q^2}\right)F_2(x, Q^2) - 2xF_1(x, Q^2) \\ &= \frac{4M^2x^2}{Q^2}F_2(x, Q^2) = \frac{M^3}{\alpha}\sum_q e_q^2 f_q(x)\frac{8x^3}{Q^2}, \end{aligned} \tag{5.51}$$

and the corresponding Compton amplitude is

$$T_L(x, Q^2) = \left(1 + \frac{\nu^2}{Q^2}\right)T_2(x, Q^2) - T_1(x, Q^2) = T_2(x, Q^2). \qquad (5.52)$$

The remarkable feature of the parton model is the Callan–Gross relation [22],

$$F_2(x, Q^2) = 2xF_1(x, Q^2), \qquad (5.53)$$

or, equivalently, in terms of the cross sections,

$$\sigma_T = \frac{\nu^2}{Q^2}\sigma_L = \frac{Q^2}{4M^2x^2}\sigma_L. \qquad (5.54)$$

This relation immediately implies the convergence of the unsubtracted dispersion relation for T_L, as long as the unsubtracted relation is valid for T_2.

Writing the unsubtracted dispersion relations for the Compton amplitudes in the parton model,

$$T_1(x, Q^2) = \frac{2\alpha}{M}\int_0^1 \frac{d\zeta}{\zeta^3}\frac{x^2F_1(\zeta, Q^2)}{\left(\frac{x}{\zeta}\right)^2 - 1 - i\epsilon}, \qquad (5.55)$$

$$T_2(x, Q^2) = \frac{4M\alpha}{Q^2}\int_0^1 \frac{d\zeta}{\zeta^2}\frac{x^2F_2(\zeta, Q^2)}{\left(\frac{x}{\zeta}\right)^2 - 1 - i\epsilon}, \qquad (5.56)$$

one can deduce that the amplitude T_2 (and, consequently, T_L) satisfies the unsubtracted dispersion relation, but T_1 does not. The latter, however, satisfies the once-subtracted dispersion relation with a vanishing subtraction function. This fact together with the Callan–Gross relation verifies the sum rule for the subtraction function (5.38). The unsubtracted dispersion relation for T_L (5.35), and the relations for T_1 following from it, hold in the naïve parton model.

5.9 Further spin-dependent relations

So far, we have considered the spin-dependent amplitude S_2 in the limit $\nu \to 0$, and S_1 in the combined soft-photon limit of both ν and $Q^2 \to 0$. It may be worthwhile to consider the known low-energy expansions of these amplitudes [23, 24]:

$$\begin{aligned} S_1^{\rm np}(\nu, Q^2) = &\frac{2\pi\alpha}{M}\varkappa^2\left[-1 + \frac{1}{3}Q^2\langle r^2\rangle_2\right] + 4\pi M\gamma_0\,\nu^2 \\ &+ 4\pi M\, Q^2\{\gamma_{E1M2} - 3M\alpha[P'^{(M1,\,M1)1}(0) + P'^{(L1,\,L1)1}(0)]\} \\ &+ O(\nu^4, \nu^2Q^2, Q^4), \end{aligned} \qquad (5.57a)$$

$$\nu S_2^{\text{np}}(\nu, 0) = -\,4\pi M^2\nu^2\{\gamma_0 + \gamma_{E1E1} - 3M\alpha[P^{'(M1,\,M1)1}(0) - P^{'(L1,\,L1)1}(0)]\} + O(\nu^4), \tag{5.57b}$$

where $\langle r^2\rangle_2 = -(6/\varkappa)\, d/dQ^2\, F_2(Q^2)|_{Q^2=0}$ is the mean-square Pauli radius; γ_{E1M2} and γ_{E1E1} are the spin polarizabilities of the corresponding multipolarities; γ_0 is the forward spin polarizability; and Ps are the generalized polarizabilities (GPs), see equation (5.64) below. In case of S_1, the first term originates from the difference between the Born contribution, whereas polarizabilities come from the non-Born part of the amplitudes.

On the right-hand side of the dispersion relations (5.8c) and (5.8d) we have an expansion in terms of moments of the spin structure functions g_1 and g_2. The first moment of g_1 is related to the generalized GDH integrals:

$$I_1(Q^2) = \frac{2M^2}{Q^2}\int_0^{x_0} dx\, g_1(x,\, Q^2), \tag{5.58a}$$

$$I_A(Q^2) = \frac{2M^2}{Q^2}\int_0^{x_0} dx\, g_{\text{TT}}(x,\, Q^2) = \frac{M^2}{4\pi^2\alpha}\int_{\nu_0}^{\infty} \frac{d\nu}{\nu}\sigma_{\text{TT}}(\nu,\, Q^2), \tag{5.58b}$$

where

$$g_{\text{TT}}(x,\, Q^2) = g_1(x,\, Q^2) - \frac{4M^2x^2}{Q^2}g_2(x,\, Q^2). \tag{5.59}$$

In the real-photon limit, $Q^2 \to 0$, these integrals turn into the GDH sum rule:

$$-\frac{1}{4}\varkappa^2 = I_1(0) = I_A(0). \tag{5.60}$$

The second moments appear in the following generalization of the forward spin polarizabilities [13]:

$$\gamma_0(Q^2) = \frac{16\alpha M^2}{Q^6}\int_0^{x_0} dx\, x^2\, g_{\text{TT}}(x,\, Q^2) = \frac{1}{2\pi^2}\int_0^{\infty} \frac{d\nu}{\nu^3}\sigma_{\text{TT}}(\nu,\, Q^2), \tag{5.61a}$$

$$\delta_{\text{LT}}\,(Q^2) = \frac{16\alpha M^2}{Q^6}\int_0^{x_0} dx\, x^2[g_1 + g_2](x,\, Q^2) = \frac{1}{2\pi^2}\int_0^{\infty} \frac{d\nu}{\nu^2 Q}\sigma_{\text{LT}}(\nu,\, Q^2), \tag{5.61b}$$

which evidently satisfy the following relations at $Q^2 = 0$:

$$\gamma_0 = \lim_{Q^2\to 0}\frac{16\alpha M^2}{Q^6}\int_0^{x_0} dx\, x^2\, g_1(x,\, Q^2), \tag{5.62a}$$

$$\delta_{\text{LT}} = \gamma_0 + \lim_{Q^2\to 0}\frac{16\alpha M^2}{Q^6}\int_0^{x_0} dx\, x^2\, g_2\,(x,\, Q^2). \tag{5.62b}$$

We recognize that the first of these is simply the FSP sum rule (4.19) written in terms of the spin structure function.

From the Q^2 term in the expansion of S_1, and the ν^2 term in the expansion of S_2, one obtains, respectively, the following relations [24]:

$$I_1^{'}(0) = \frac{1}{12}\varkappa^2\langle r^2\rangle_2 + \frac{1}{2\alpha}M^2\gamma_{\mathrm{E1M2}} - \frac{3}{2}M^3\,[P^{'(\mathrm{M1,\,M1})1}(0) + P^{'(\mathrm{L1,\,L1})1}(0)], \quad (5.63a)$$

$$\delta_{\mathrm{LT}} = -\gamma_{\mathrm{E1E1}} + 3\alpha M\,[P^{'(\mathrm{M1,\,M1})1}(0) - P^{'(\mathrm{L1,\,L1})1}(0)], \quad (5.63b)$$

where the momentum derivatives of the GPs are given by

$$P^{'(\mathrm{M1,\,M1})1}(0) \pm P^{'(\mathrm{L1,\,L1})1}(0) \equiv \frac{d}{dQ^2}[P^{(\mathrm{M1,\,M1})1}(Q^2) \pm P^{(\mathrm{L1,\,L1})1}(Q^2)]_{Q^2=0}, \quad (5.64)$$

with the superscript indicating the multipolarities, $L1(M1)$ denote the electric (magnetic) dipole transitions of the initial and final photons, and '1' implies that these transitions involve the spin-flip of the nucleon, see [25, 26].

As we can see, the higher-order relations are quickly becoming very elaborate. They still are useful in relating the various polarizabilities and moments of structure functions, but this is as far as we go here. The persistent reader might be interested to see further discussion of these relations in [24, 27, 28].

References

[1] Low F E 1954 Scattering of light of very low frequency by systems of spin 1/2 *Phys. Rev.* **96** 1428–32

[2] Gell-Mann M and Goldberger M L 1954 Scattering of low-energy photons by particles of spin 1/2 *Phys. Rev.* **96** 1433–8

[3] Burkhardt H and Cottingham W N 1970 Sum rules for forward virtual Compton scattering *Ann. Phys.* **56** 453–63

[4] Schwinger J S 1975 Source theory viewpoints in deep inelastic scattering *Proc. Natl Acad. Sci.* **72** 1–5

Schwinger J 1975 Source theory viewpoints in deep inelastic scattering *Electromagnetic Interactions and Field Theory Acta Phys. Austriaca Suppl.* 14 (Berlin: Springer) pp 471–88

[5] Schwinger J 1975 Source theory discussion of deep inelastic scattering with polarized particles *Proc. Natl Acad. Sci. USA* **72** 1559–63

[6] Rashid H 1976 A simple derivation of Schwinger's sum rule for spin dependent structure functions *Nuovo Cim.* A **33** 447

[7] Hagelstein F and Pascalutsa V 2018 Dissecting the hadronic contributions to $(g-2)_\mu$ by Schwingers sum rule *Phys. Rev. Lett.* **120** 072002

[8] Schwinger J 1949 Quantum electrodynamics. II. Vacuum polarization and self-energy *Phys. Rev.* **75** 651–79

[9] Tsai W-y, DeRaad L L and Milton K A 1975 Verification of virtual Compton-scattering sum rules in quantum electrodynamics *Phys. Rev.* D **11** 3537–40

[10] Liang Y, Christy M, Ent R and Keppel C 2006 Q^2 evolution of generalized Baldin sum rule for the proton *Phys. Rev.* C **73** 065201

[11] Sibirtsev A and Blunden P G 2013 Q^2 evolution of the electric and magnetic polarizabilities of the proton *Phys. Rev.* C **88** 065202

[12] Hall N L, Thomas A W and Young R D 2014 Momentum transfer dependence of the proton's electric and magnetic polarizabilities *Phys. Rev.* D **89** 117502

[13] Drechsel D, Pasquini B and Vanderhaeghen M 2003 Dispersion relations in real and virtual Compton scattering *Phys. Rep.* **378** 99–205

[14] Hagelstein F, Miskimen R and Pascalutsa V 2016 Nucleon polarizabilities: from Compton scattering to hydrogen atom *Prog. Part. Nucl. Phys.* **88** 29–97

[15] Siegert A J F 1937 Note on the interaction between nuclei and electromagnetic radiation *Phys. Rev.* **52** 787–9

[16] Bernabeu J and Tarrach R 1975 Unsubtracted dispersion relation for the longitudinal Compton amplitude? *Phys. Lett.* B **55** 183

[17] L'vov A I 1998 Electric polarizability of nuclei and a longitudinal sum rule *Nucl. Phys.* A **638** 756–64

[18] Gasser J, Hoferichter M, Leutwyler H and Rusetsky A 2015 Cottingham formula and nucleon polarisabilities *Eur. Phys. J.* C **75** 375
Gasser J, Hoferichter M, Leutwyler H and Rusetsky A 2020 Erratum to: Cottingham formula and nucleon polarisabilities *Eur. Phys. J.* C **80** 353

[19] Bernabeu J, Gomez Dumm D and Orlandini G 1998 Electric polarizability of nuclei from a longitudinal sum rule *Nucl. Phys.* A **634** 463–71

[20] Llanta E and Tarrach R 1978 Polarizability sum rules in QED *Phys. Lett.* B **78** 586–8

[21] Peskin M E and Schroeder D V 1995 *An Introduction to Quantum Field Theory* (Reading, MA: Addison-Wesley)

[22] Callan C G Jr and Gross D J 1969 High-energy electroproduction and the constitution of the electric current *Phys. Rev. Lett.* **22** 156–9

[23] Drechsel D *et al* 1998 Low-energy and low-momentum representation of the virtual Compton scattering amplitude *Phys. Rev.* C **58** 1751–7

[24] Pascalutsa V and Vanderhaeghen M 2015 Polarizability relations across real and virtual Compton scattering processes *Phys. Rev.* D **91** 051503(R)

[25] Guichon P A M, Liu G Q and Thomas A W 1995 Virtual Compton scattering and generalized polarizabilities of the proton *Nucl. Phys.* A **591** 606–38

[26] Guichon P A M and Vanderhaeghen M 1998 Virtual Compton scattering off the nucleon *Prog. Part. Nucl. Phys.* **41** 125–90

[27] Lensky V, Pascalutsa V, Vanderhaeghen M and Kao C 2017 Spin-dependent sum rules connecting real and virtual Compton scattering verified *Phys. Rev.* D **95** 074001

[28] Pasquini B and Vanderhaeghen M 2018 Dispersion theory in electromagnetic interactions *Ann. Rev. Nucl. Part. Sci.* **68** 75–103

Chapter 6

Sum rules for light-by-light scattering

The light-by-light (LbL) scattering is a well-known phenomenon of purely quantum origin [1–4]. Recently it has been observed in ultra-peripheral heavy-ion collisions [5]. Indirectly, however, LbL scattering plays an evident role through the quantum (loop) effects. It is known in particular for its influence on precision observables, such as $g-2$ of the electron and muon, the hydrogen Lamb shift, etc. In this chapter, we again exploit the general constraints of causality and unitarity in order to express the forward (and, equally, the backward) LbL scattering amplitude in terms of the integrals of $\gamma\gamma$ fusion cross sections. Using an effective field theory (EFT) approach, we derive the 'soft-photon theorems' and subsequently the LbL sum rules.

6.1 Compton scattering off a photon

Those familiar with the Compton scattering sum rules, seen in the previous chapters, will be able to grasp the LbL scattering case by viewing it as a (very) special example of Compton scattering. The LbL is of course a Compton scattering off a massless neutral spin-1 target—the photon. The main difference is that the probe and the target are identical particles, and so there is an additional symmetry: the forward and backward scattering are identical. This symmetry is encompassed in an additional crossing symmetry, as will be seen below.

Moreover, the photon as the target is not only neutral, it can have no electromagnetic moments whatsoever. This is a direct consequence of the charge conjugation symmetry, and, more specifically, *Furry's theorem*. Therefore, for example, the analogue of the GDH sum rule will have a strict 0 on the left-hand side, rather than an anomalous magnetic moment. Nevertheless, the general strategy of the derivation is the same.

6.2 Symmetries, unitarity, and dispersion relations

As in the Compton case, we shall only use the general assumptions about analyticity, crossing symmetry, unitarity, and the low-energy limit of the forward elastic

doi:10.1088/978-0-7503-3431-0ch6

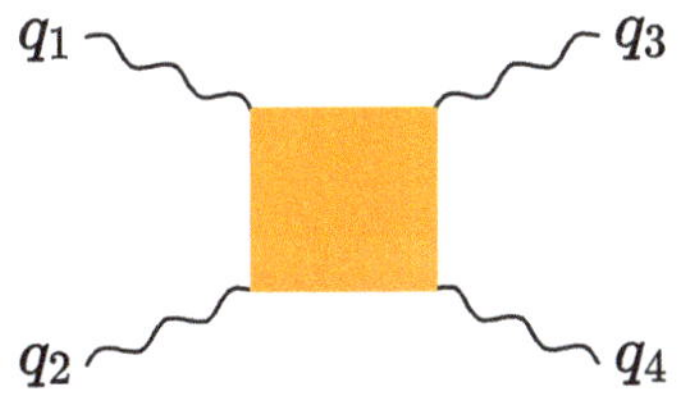

Figure 6.1. A generic diagram of light-by-light scattering, indicating the four-momenta of the photons; q_1, q_2—incoming, and q_3, q_4—outgoing.

scattering amplitude. Only now, it is the amplitude of the LbL process: $\gamma\gamma \to \gamma\gamma$, see figure 6.1. Denoting the corresponding Feynman and helicity amplitudes as, respectively, $\mathcal{M}$ and M, their relation is

$$M_{\lambda_1\lambda_2\lambda_3\lambda_4} = \varepsilon_{\lambda_4}^{*\mu_4}(\vec{q}_4)\, \varepsilon_{\lambda_3}^{*\mu_3}(\vec{q}_3)\, \varepsilon_{\lambda_2}^{\mu_2}(\vec{q}_2)\, \varepsilon_{\lambda_1}^{\mu_1}(\vec{q}_1) \mathcal{M}_{\mu_1\mu_2\mu_3\mu_4}, \tag{6.1}$$

where $\varepsilon(\vec{q})$ are the photon polarization four-vectors and λ is the helicity ($\lambda = \pm$ for the right- and left-handed circular polarization, respectively). For real photons traveling along the z axis, i.e. with $\vec{q} = (0, 0, \omega)$, the polarization vectors are $\varepsilon_\lambda(\pm\vec{q}) = \frac{1}{\sqrt{2}}(0, \mp\lambda, -i, 0)$.

For q_i the photon four-momenta defined in figure 6.1, the Mandelstam variables are:

$$s = (q_1 + q_2)^2 = (q_3 + q_4)^2 = 4\omega^2, \tag{6.2a}$$

$$t = (q_1 - q_3)^2 = (q_2 - q_4)^2 = -2\omega^2(1 - \cos\vartheta), \tag{6.2b}$$

$$u = (q_1 - q_4)^2 = (q_2 - q_3)^2 = -s - t, \tag{6.2c}$$

where ω and ϑ are the photon energy and scattering angle in the 'collider frame', defined by $\vec{q}_1 + \vec{q}_2 = \vec{q}_3 + \vec{q}_4 = 0$. The latter condition is equivalent to the center-of-mass frame, but since there is no mass here, this name seems dubious.

In the forward kinematics $q_3 = q_1$, $q_4 = q_2$, and hence $t = 0 = \vartheta$, $u = -s$. The general Lorentz structure of the Feynman amplitude is then given by

$$\mathcal{M}_{\mu_1\mu_2\mu_3\mu_4} = A(s)\, g_{\mu_4\mu_2}\, g_{\mu_3\mu_1} + B(s)\, g_{\mu_4\mu_1}\, g_{\mu_3\mu_2} + C(s)\, g_{\mu_4\mu_3}\, g_{\mu_2\mu_1}, \tag{6.3}$$

where $g_{\mu\nu}$ is the Minkowski metric tensor, and A, B, and C are the scalar amplitudes. Crossing symmetry under $1 \leftrightarrow 3$ ($s \leftrightarrow u$) requires

$$\mathcal{M}_{\mu_1\mu_2\mu_3\mu_4} = A(u)\, g_{\mu_4\mu_2}\, g_{\mu_3\mu_1} + B(u)\, g_{\mu_4\mu_3}\, g_{\mu_1\mu_2} + C(u)\, g_{\mu_1\mu_4}\, g_{\mu_3\mu_2}, \tag{6.4}$$

and hence, matching the two expressions, we have $A(-s) = A(s)$, $B(-s) = C(s)$. As a result, there are three independent nonvanishing helicity amplitudes,

$$\begin{aligned} M_{++++}(s) &= A(s) + C(s), \\ M_{+-+-}(s) &= A(s) + B(s), \\ M_{++--}(s) &= B(s) + C(s), \end{aligned} \tag{6.5}$$

satisfying the following crossing relations:

$$M_{+-+-}(s) = M_{++++}(-s), \qquad M_{++--}(s) = M_{++--}(-s). \tag{6.6}$$

So far everything is just as in the Compton process, with the difference that there are three helicity amplitudes instead of four. Recall that our general counting rules of section 4.1 give $2S + 2$ amplitudes for a target with integer spin S. Indeed, for a massive spin-1 target we would have in addition the amplitude M_{+0+0}, with $\lambda = 0$ representing the longitudinal polarization. Since the photon is massless, we have one amplitude fewer. Although we will have to revisit this argument for the virtual photon considered in the next section.

Let us briefly remark on another unique property of the LbL scattering—the symmetry between the forward and backward kinematics. How does one distinguish whether the photons pass forwards or scatter backwards? One cannot, hence it is a symmetry. In fact, it is the $3 \leftrightarrow 4$ crossing symmetry, which interchanges t and u. As the result, the number of possible sum rules doubles. For example, in addition to the analogue of the Baldin sum rule, there is a sum rule for another combination of the structure constants, see equation (6.21) below.

The next step in the derivation is the optical theorem which relates the imaginary (also called 'absorptive') part of these amplitudes to the photoabsorption cross sections, which in this case are called the $\gamma\gamma$-fusion cross sections. Obviously, the $\gamma\gamma$ state '++' has the total helicity 0 (because the photons are oriented in the opposite directions), whereas the '+−' has the total helicity 2. Therefore, in a certain convention for normalization, we have

$$\mathrm{Im}\, M_{++++}(s) = \frac{1}{2} s\, \sigma_0(s), \qquad \mathrm{Im}\, M_{+-+-}(s) = \frac{1}{2} s\, \sigma_2(s). \tag{6.7}$$

The case of the double-helicity flip amplitude M_{++--} is more tricky. To form an appropriate cross section, we need to envision which combination of the squared $\gamma\gamma$-fusion amplitudes, denoted as $T_{\lambda_1\lambda_2}$ will give: $T_{++}T_{--}$. It turns out, we need linearly polarized amplitudes. The linear polarizations are simply given as

$$\varepsilon_x = (0,\, 1,\, 0,\, 0) = \frac{1}{\sqrt{2}}(\varepsilon_+ - \varepsilon_-), \tag{6.8a}$$

$$\varepsilon_y = (0,\, 0,\, 1,\, 0) = \frac{i}{\sqrt{2}}(\varepsilon_+ + \varepsilon_-), \tag{6.8b}$$

and therefore the linearly polarized cross sections are expressed in terms of the circularly polarized ones as follows:

$$\sigma_{\parallel} \sim |T_{xx}|^2 = \frac{1}{4}\,|\, T_{++} - T_{+-} - T_{-+} + T_{--}\,|^2, \tag{6.9a}$$

$$\sigma_{\perp} \sim |T_{xy}|^2 = \frac{1}{4}\,|\, T_{++} + T_{+-} - T_{-+} - T_{--}\,|^2, \tag{6.9b}$$

$$\sigma_{||} - \sigma_{\perp} \sim \mathrm{Re}\,(T_{++}T^{*}_{--} + T_{-+}T^{*}_{+-} - T_{++}T^{*}_{+-} - T_{-+}T^{*}_{--}). \tag{6.9c}$$

In the difference, only the first term survives, because the rest are products of amplitudes which have different helicities. We therefore arrive at the following statement of unitarity:

$$\mathrm{Im}\, M_{++--}(s) = \frac{1}{2}s\left[\,\sigma_{||}(s) - \sigma_{\perp}(s)\,\right]. \tag{6.10}$$

We finally turn to the causality constraint, and would like to express it in terms of dispersion relations. To do so, we first form the amplitudes which have a definite sign under the reflection of s:

$$f^{(\pm)}(s) = M_{++++}(s) \pm M_{+-+-}(s), \tag{6.11a}$$

$$g(s) = M_{++--}(s), \tag{6.11b}$$

such that $f^{(\pm)}(-s) = \pm f^{(\pm)}(s)$ and $g(-s) = g(s)$, and assume they are analytic functions of s^2 everywhere in the complex plane except the branch cuts along the positive real axis. The resulting dispersion relations are

$$\mathrm{Re}\begin{Bmatrix} f^{(+)}(s) \\ g(s) \end{Bmatrix} = \frac{2}{\pi}\fint_0^{\infty}\frac{ds'\, s'}{s'^2 - s^2}\,\mathrm{Im}\begin{Bmatrix} f^{(+)}(s') \\ g(s') \end{Bmatrix}, \tag{6.12a}$$

$$\mathrm{Re} f^{(-)}(s) = \frac{2s}{\pi}\fint_0^{\infty} ds'\,\frac{\mathrm{Im} f^{(-)}(s')}{s'^2 - s^2}. \tag{6.12b}$$

Substituting the expressions for the imaginary parts as dictated by the optical theorem, we obtain

$$\mathrm{Re} f^{(+)}(s) = \frac{2}{\pi}\fint_0^{\infty} ds'\, s'^2\,\frac{\sigma(s')}{s'^2 - s^2}, \tag{6.13a}$$

$$\mathrm{Re} f^{(-)}(s) = \frac{s}{\pi}\fint_0^{\infty} ds'\,\frac{s'\,\Delta\sigma(s')}{s'^2 - s^2}, \tag{6.13b}$$

$$\mathrm{Re}\, g(s) = \frac{1}{\pi}\fint_0^{\infty} ds'\, s'^2\,\frac{\sigma_{||}(s') - \sigma_{\perp}(s')}{s'^2 - s^2}, \tag{6.13c}$$

where $\sigma = 1/2(\sigma_0 + \sigma_2) = 1/2(\sigma_{||} + \sigma_{\perp})$ is the unpolarized total cross section, and $\Delta\sigma = \sigma_2 - \sigma_0$ is the helicity-difference cross section. In order to consider the low-energy expansion of these expressions, we must know something about the low-energy behavior of the LbL amplitudes. In other word, we need an analogue of the soft-photon theorem. For this, we shall make a quick detour into the EFT approach to the low-energy expansion.

6.3 Effective field theorems

The most appropriate description of photons is of course given by electrodynamics, which is a field theory, and so there is no better way to gain insight into the low-energy LbL interaction than the field theory itself.

Let us recall that the free photon propagation follows from the well-known Lagrangian density:

$$\mathcal{L}_{\text{free}} = -\frac{1}{4}F_{\mu\nu}F^{\mu\nu}, \qquad F_{\mu\nu} = \partial_\mu A_\nu - \partial_\nu A_\mu, \tag{6.14}$$

where $A_\mu(x)$ is the electromagnetic vector field describing the photon. It is invariant under the gauge transformations, $A_\mu(x) \to A_\mu(x) + \partial_\mu\varphi(x)$, where φ is an arbitrary scalar field. Maintaining this gauge symmetry is very important for the correct degree-of-freedom content of the theory. Without it, the vector field would have up to four independent spin degrees of freedom, or polarizations, rather than two as is appropriate for a massless particle with spin. Therefore, we need to maintain this symmetry when constructing the possible interactions, and the only way to maintain it in the absence of other fields is to construct the Lagrangian in terms of $F_{\mu\nu}$, which is an explicitly gauge-invariant object. Another sensible requirement is that the couplings of the electromagnetic field are symmetric under parity and time-reversal, since those discrete symmetries are known to be generally preserved in the electromagnetic interactions.

The simplest coupling containing only the electromagnetic field is therefore given by the following Euler–Heisenberg Lagrangian [2],

$$\mathcal{L}_{\text{EH}} = c_1\left(F_{\mu\nu}F^{\mu\nu}\right)^2 + c_2\left(F_{\mu\nu}\tilde{F}^{\mu\nu}\right)^2, \tag{6.15}$$

where $\tilde{F}^{\mu\nu} = \varepsilon^{\mu\nu\alpha\beta}\partial_\alpha A_\beta$ is the dual field-strength tensor, and c_i are arbitrary coupling constants.

One cannot have cubic couplings, or any odd number of $F_{\mu\nu}$ for that matter, not only because it is forbidden by Furry's theorem, but also because such couplings are trivially 0. For example, writing a term like $(g_{\alpha\beta}F^{\mu\alpha}F^{\beta\nu})F_{\mu\nu}$, we notice the object in the brackets is symmetric in μ and ν, while $F_{\mu\nu}$ is of course antisymmetric; hence, their contraction is 0. It is a simple exercise to show that this holds in general for any odd number of $F_{\mu\nu}$.

The above quartic coupling are the simplest, but not the only ones. One can hang a derivative on $F_{\mu\nu}$ and then have, for example

$$\mathcal{L}^{(10)} = c_3\left(\partial_\alpha F_{\mu\nu}\right)\left(\partial^\alpha \tilde{F}^{\mu\nu}\right)\left(F_{\mu'\nu'}\tilde{F}^{\mu'\nu'}\right) + c_4\left(\partial_\alpha F_{\mu\nu}\right)\left(\partial^\alpha F^{\mu\nu}\right)\left(F_{\mu'\nu'}F^{\mu'\nu'}\right) + \cdots \tag{6.16}$$

These couplings will, however, play a lesser role at low energies, because each derivative translates into the photon four-momentum and, hence, a higher power of energy. In general, to estimate how important a given coupling is one can just count its dimension—the smaller the dimension the more important it is at low energies. For instance, the dimension of the Euler–Heisenberg couplings is 8 (1 for each field

A_μ and 1 for each derivative), whereas the dimension of the higher-derivative couplings in (6.16) is indicated by the superscript on the left-hand side (the complete dimension-10 quartic coupling is discussed in [6]).

To conclude this part, the Euler–Heisenberg Lagrangian describes the leading-order photon self-couplings at low energies. The soft-photon theorem for the LbL scattering amplitudes simply follows from it:

$$\begin{Bmatrix} f^{(+)}(s) \\ g(s) \end{Bmatrix} = 8(c_1 \pm c_2)s^2 + O(s^4), \qquad f^{(-)}(s) = O(s^3). \tag{6.17}$$

6.4 The sum rules

And so, expanding the left-hand and right-hand sides of equation (6.13a) in powers of s and matching them at each order yields a number of sum rules. At the zeroth order in s, there are no contributions on the left-hand side of (6.13a) and (6.13c), and one finds

$$0 = \int_0^\infty ds \left[\sigma_{||}(s) \pm \sigma_\perp(s)\right]. \tag{6.18}$$

This certainly cannot work for '+', since the unpolarized cross section is a positive-definite quantity. Moreover, empirically the unpolarized cross section for $\gamma\gamma$ fusion is slowly rising at large s and the integral diverges. We have seen this before, for the failed Thomson-term sum rule in Compton scattering. Here, the Thomson term is absent, and we have one more sum rule for the '−' case. The latter is the consequence of the above-mentioned forward-backward symmetry of the $\gamma\gamma$ system. In either case, however, we have to scrap these sum rules as violating the assumption of an unsubtracted dispersion relation.

At the first order in s, the relation (6.13b) yields a valid sum rule:

$$\boxed{0 = \int_0^\infty ds\, \frac{\sigma_2(s) - \sigma_0(s)}{s}}. \tag{6.19}$$

First written down by Gerasimov and Moulin [7], this sum rule is fully analogous to the GDH sum rule. In fact, Brodsky and Schmidt [8] simply infer this sum rule from the GDH sum rule by arguing that the photon has no anomalous magnetic moment due to Furry's theorem. The latter property makes this sum rule even more profound, because it allows for a remarkable generalization to the case when one of the photons is virtual (with virtuality $q^2 < 0$):

$$\boxed{0 = \int_0^\infty ds\, \frac{\sigma_2(s, q^2) - \sigma_0(s, q^2)}{s - q^2}}, \tag{6.20}$$

where the cross sections are the ones of $\gamma\gamma^*$ fusion. This sum rule can be cast as a sum rule for the photon spin structure function g_1^γ [9–12], i.e. $\int_0^1 dx\, g_1^\gamma(x, Q^2) = 0$. It has no analogue in Compton scattering.

Going further, at order s^2 we finally have a non-trivial contribution on the left-hand side of the relations (6.13a), (6.13c), and we have the following two sum rules [13]:

$$\boxed{c_1 \pm c_2 = \frac{1}{8\pi}\int_0^\infty ds\, \frac{\sigma_{||}(s) \pm \sigma_{\perp}(s)}{s^2}}. \tag{6.21}$$

The first one, with the '+', is an analogue of the Baldin sum rule for the forward combination of the scalar polarizabilities, where the structure constants c_i play the role of 'photon polarizabilities'. Taking the '−', we find the other combination of polarizabilities, the backward-scattering one. The latter sum rule is a consequence of the additional symmetry between the backward and forward scattering, and therefore has no analogue in Compton scattering.

Both structure constants can thus be determined from the (polarized) $\gamma\gamma$-fusion cross sections. Even without knowing anything about these cross sections, we may conclude that these constants are positive-definite: $c_1, c_2 > 0$, and as a result, the low-energy interaction between two photons is always attractive. (In case the latter statement seem discontinued from the former, it should be helpful to see that the positivity of the Euler–Heisenberg Lagrangian (6.15) translates into a negative potential energy, hence the attractive potential.)

Certainly, one can keep going to higher orders in s and/or consider the situation when one or more photons are virtual. There has indeed been some work done in these directions recently, see e.g. [6]. Here, however, we are done with derivations. In the remainder of this chapter we consider some implications of the few simple sum rules which we have just derived.

6.5 Perturbative verification

It is often useful to check how the sum rules work in a simple theory such as QED. There, the leading process of $\gamma\gamma$ fusion is the pair production of charged particles, see figures 3.1 and 6.2. The particles can be scalars (spin-0), Dirac fermions (spin-1/2), or massive vector bosons (spin-1). The superconvergence sum rule (6.20) has been verified for all of these cases [6]. For example, the spin-1/2 pair-production helicity-difference cross section in $\gamma^*\gamma$ collisions is given by

$$\Delta\sigma^{(e^+e^-)}(s, q^2) = \frac{8\pi\alpha^2}{(s-q^2)^2}\left\{-(3s+q^2)\sqrt{1-\frac{4m^2}{s}} + 2(s+q^2)\,\mathrm{arctanh}\sqrt{1-\frac{4m^2}{s}}\right\}, \tag{6.22}$$

where q^2 is the time-like virtuality of one of the photons, the other one is real; m is the mass of the particle, so that the threshold is at $s_0 = 4m^2$. This cross section is

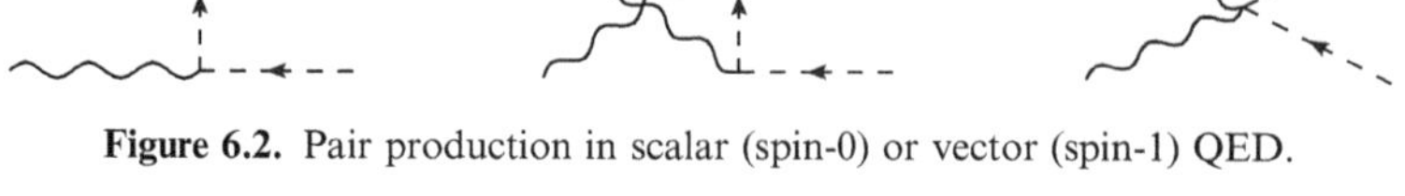

Figure 6.2. Pair production in scalar (spin-0) or vector (spin-1) QED.

plotted in figure 6.3 for three different values of q^2. One sees that in all cases the low- and high-energy contributions cancel, and the sum rule is verified for any $q^2 < 4m^2$.

Computing the linearly polarized cross sections, and using the sum rules (6.21), we obtain the values of the low-energy constants shown in table 6.1. One can see how they depend on the spin of the matter particles. There is a clear pattern for the difference of these constants which allows us to surmise the expression for any spin, see the last column.

Let us note that the spin-1 result is in agreement with the one-loop electroweak correction to LbL scattering (figure 6.4), which can be obtained by the low-energy

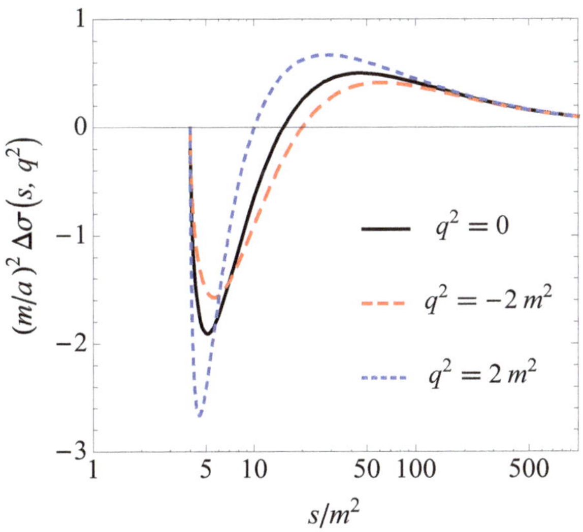

Figure 6.3. Helicity-difference cross section ($\Delta\sigma = \sigma_2 - \sigma_0$) of $\gamma^*\gamma \to e^+e^-$ in QED at leading order (corresponds with the diagrams in figure 3.1) for different photon virtualities.

Table 6.1. Low-energy constants of the Euler–Heisenberg Lagrangian in QED. The last column shows the ratio of the difference for a given spin over the difference for spin-0. The expression for arbitrary S is a conjecture.

	c_1	c_2	$\frac{c_1 - c_2}{(c_1 - c_2)_0}$
Spin 0	$\frac{7}{1440}\frac{\alpha^2}{m^4}$	$\frac{1}{1440}\frac{\alpha^2}{m^4}$	1
Spin 1/2	$\frac{1}{90}\frac{\alpha^2}{m^4}$	$\frac{7}{360}\frac{\alpha^2}{m^4}$	−2
Spin 1	$\frac{29}{160}\frac{\alpha^2}{m^4}$	$\frac{27}{160}\frac{\alpha^2}{m^4}$	3
Spin S			$(-1)^{2S}(2S + 1)$

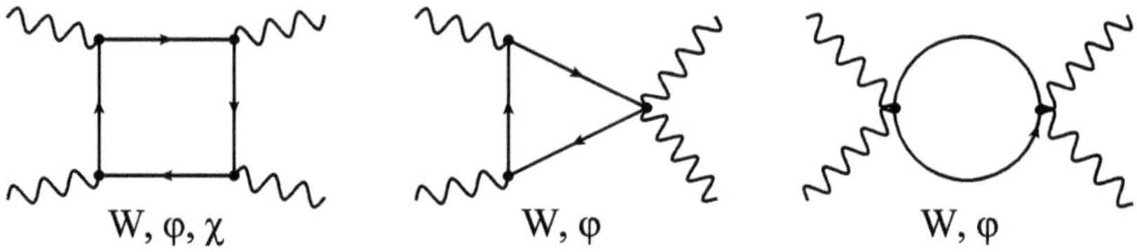

Figure 6.4. One-loop Feynman graphs in the standard electroweak theory. Here W stands for the charged gauge bosons, φ is the Higgs sector and χ denotes Faddeev–Popov ghosts.

expansion of the expressions from Bohm and Schuster [14]. It is interesting that to obtain this result in the electroweak theory one needs to take care of the Higgs mechanism and the ghosts, whereas the sum rule calculation is simpler: the tree-level production of massive vector bosons (figure 6.2), and the dispersion integral of (6.21).

We can go beyond the tree level. The simplest example is provided by scalar QED extended by the self-coupling of the scalar field (i.e. the ϕ^4 theory), described by the following Lagrangian density:

$$\mathcal{L} = (D^\mu\phi)^* D_\mu\phi - m^2\phi^*\phi + \frac{\lambda}{4}(\phi^*\phi)^2, \tag{6.23}$$

where $\phi(x)$ is a complex scalar field, λ is its self-coupling constant, and $D_\mu = \partial_\mu + i.\ e.\ A_\mu$ is the covariant derivative. Denoting the helicity amplitudes for the process $\gamma\gamma \to \phi\phi^*$ by T_{++} and T_{+-}, the cross sections for total helicity 0 and 2 are found as

$$\sigma_0(s) = \frac{v}{32\pi s}\int_{-1}^{1} d\cos\theta \mid T_{++}(s,\theta)|^2, \tag{6.24}$$

$$\sigma_2(s) = \frac{v}{32\pi s}\int_{-1}^{1} d\cos\theta \mid T_{+-}(s,\theta)|^2, \tag{6.25}$$

where θ is the angle of one of the particles with respect to the z axis in the collider frame, and $v = \sqrt{1 - 4m^2/s}$ is the relative velocity of the scalar's pair. To leading order in α and λ, the helicity amplitudes are [15]

$$T_{++}(s,\theta) = 8\pi\alpha\,\frac{1 - v^2}{1 - v^2\cos^2\theta} + \frac{\alpha\lambda}{2\pi}F(s), \tag{6.26a}$$

$$T_{+-}(s,\theta) = 8\pi\alpha\,\frac{v^2\sin^2\theta}{1 - v^2\cos^2\theta}, \tag{6.26b}$$

where the first term in T_{++} and the expression for T_{+-} correspond with the tree-level amplitudes for $\gamma\gamma \to \phi\bar{\phi}$ in scalar QED. The second term in T_{++} describes the one-loop production process corresponding with figure 6.5, with the gray blob denoting the four-particle amplitude evaluated to leading order in λ. Furthermore, there is a

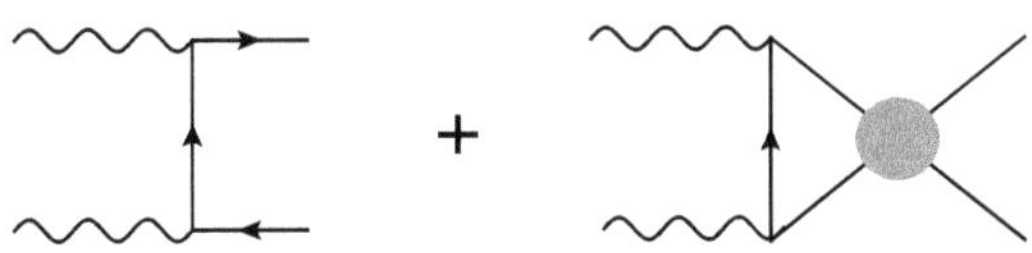

Figure 6.5. The $\gamma\gamma$-fusion to one loop in scalar QED with ϕ^4 coupling. Crossed and seagull diagrams are not shown.

form factor, describing the transition of photons to a scalar pair to one loop, given by

$$F(s) = -1 - \frac{1 - v^2}{4}\left(\ln\frac{1 + v}{1 - v} - i\pi\right)^2. \tag{6.27}$$

The corresponding result for the helicity-difference cross section, $\Delta\sigma = \sigma_2 - \sigma_0$, reads

$$\Delta\sigma(s) = \Delta\sigma^{\text{tree}}(s) - \frac{\alpha^2\lambda}{2\pi s}\,\text{Re}\,F(s)\,(1 - v^2)\,\text{arctanh}v, \tag{6.28}$$

where $\Delta\sigma^{\text{tree}}$ is the tree-level cross section in scalar QED:

$$\Delta\sigma^{\text{tree}}(s) = \frac{4\pi\alpha^2}{s}[v - 2(1 - v^2)\,\text{arctanh}\,v]. \tag{6.29}$$

As we noted above, the tree-level cross section satisfies the sum rule by itself, and we are thus left to verify the following result[1]

$$\int_{4m^2}^{\infty} ds\,\frac{\text{Re}\,F(s)}{s^3}\,\text{arctanh}\,v = 0. \tag{6.30}$$

6.6 Non-perturbative verification: bound state

It is not a surprise that the sum rule (6.19) works out in the above one-loop example. What is much more intriguing is that it works even when the self-interactions are summed into a 'bubble chain', see figure 6.6. The previous calculation was done to first order in λ, which corresponds to replacing the blob in figure 6.5 by only the first term in figure 6.6. However, it has been shown [15] that one can substitute any number of terms (from figure 6.6 into figure 6.5) and the sum rule still works. This is by no means trivial, because this summation is not a systematic calculation to a given order in λ. It is only including a sub-class of diagrams resulting from a Lippmann–Schwinger type of equation, which in this case, appropriate for a relativistic field theory, is the Bethe–Salpeter equation.

When keeping all of the bubbles, one obtains a non-perturbative result which gives rise to some qualitatively new features. The most remarkable one is the dynamical generation of a bound state. Namely, for some values of the coupling λ,

[1] Such checks are now as easy as pressing Shift-Enter.

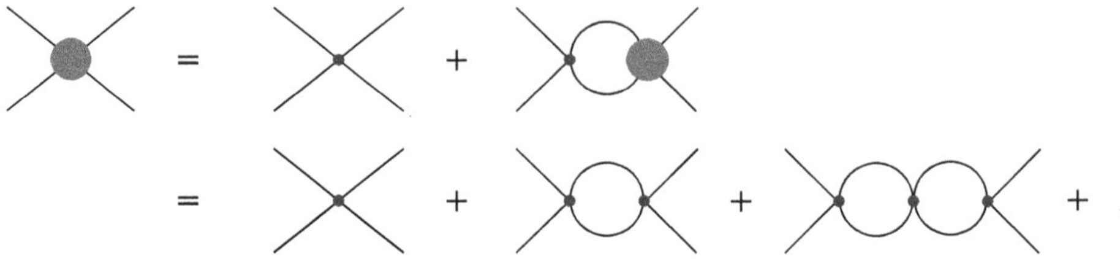

Figure 6.6. Non-perturbative summation of loops into a 'bubble chain'.

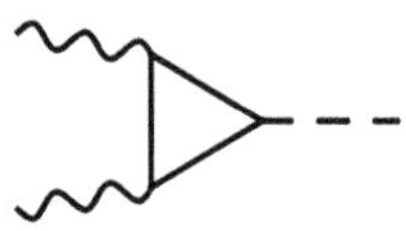

Figure 6.7. The amplitude of the bound state production in $\gamma\gamma$ fusion.

corresponding to an attractive interaction, the $\phi\phi$ rescattering amplitude acquires a pole, which is associated with a new particle Φ, a bound state of two ϕ. We do not want to go into the technical details, but it is important to realize the following.

In the presence of the bound state, the sum rule (6.19) does not work, *unless* one adds the cross section corresponding to $\gamma\gamma\rightarrow\Phi$, the production of the bound state itself, see figure 6.7. Thus the sum rule, or more generally, causality and unitarity that went into the sum rule derivation, require a dynamically generated bound state to be treated as an *asymptotic state*. It should therefore come as less of a surprise that the bound states in quantum field theory must appear as asymptotic fields, see e.g. [16, chapters 11 and 12].

Further details on the formal non-perturbative aspects of the sum rule (6.19) can be found in[15, 17], whereas we move on to consider possible empirical implications of the LbL sum rules.

6.7 Implications for mesons

The helicity-difference sum rule (6.19), along with its finite-Q^2 extension (6.20), infer an exact cancellation between the helicity 0 and 2 contributions, and as such become very useful in constraining and predicting the mechanisms of $\gamma\gamma$ fusion into hadrons. The dominant features of the $\gamma\gamma$ to multi-hadron production come firstly from the Born terms in the $\pi^+\pi^-$ (or K^+K^-) channels. These contributions are given by equation (6.29), and hence each separately obeys the sum rule. The largest contributions in the hadronic sector are thus expected to come from the resonance production: $\gamma\gamma\rightarrow M$, with M being a meson. As two real photons do not couple to a $J^P = 1^-$ or 1^+ state due to the Landau–Yang theorem, one expects the dominant contribution to come from scalar, pseudoscalar, and tensor mesons. The production of the (pseudo-)scalar mesons must then conspire with production of the tensor mesons to diminish each other's contribution in the helicity-difference integral. It will be highly non-trivial to show that this cancellation is emerging in non-perturbative QCD, but we can immediately check whether this works empirically.

One can express the $\gamma\gamma \to M$ cross section for a meson with spin J, mass m_M, and total width Γ_{tot}, using a Breit–Wigner parametrization, in terms of the decay width $\Gamma_{\gamma\gamma}$ of the meson into two photons of total helicity $J = 0, 2$, as

$$\begin{aligned} \sigma_J^{\gamma\gamma\to M}(s) &= 16\pi(2J+1)\frac{\Gamma_{\gamma\gamma}\,\Gamma_{tot}}{(s-m_M^2)^2+\Gamma_{tot}^2 m_M^2} \\ &\approx 16\pi^2(2J+1)\frac{\Gamma_{\gamma\gamma}}{m_M}\,\delta(s-m_M^2), \end{aligned} \tag{6.31}$$

where the last line is obtained in the narrow resonance approximation, which will be used in what follows. The contribution of an individual meson to the sum rule,

$$\int_0^\infty ds\,\frac{\Delta\sigma^{\gamma\gamma\to M}(s)}{s} \approx \frac{16\pi^2\Gamma_{\gamma\gamma}}{m_M^3}\times\begin{cases}-1, & \text{for } J=0\\ 5, & \text{for } J=2\end{cases}, \tag{6.32}$$

decreases rapidly with mass, and hence the lightest mesons, such as the pseudoscalars π^0, η and η', should be considered first. The corresponding results are shown in table 6.2. Note that the pion contribution is simply given by the chiral anomaly as $-\alpha^2/(4\pi f_\pi^2)$, with $f_\pi = 92.4$ MeV the pion decay constant.

The large negative contribution from pseudoscalar mesons must be compensated by an equivalent strength in the helicity-2 cross section, $\sigma_2(s)$. The dominant feature of the helicity-two cross section in the resonance region arises from the multiplet of tensor mesons $f_2(1270)$, $a_2(1320)$, and $f_2'(1525)$. In table 6.2, we can see that the sum rule contribution of the lowest isovector tensor meson composed of light quarks, $a_2(1320)$, compensates to around 70% the π^0 contribution governed by the chiral anomaly. For the isoscalar states composed of light quarks, the cancellation is even more remarkable. The sum of $f_2(1270)$ and $f_2'(1525)$ cancels nearly exactly the combined contribution of the η and η' (see the last two rows of table 6.2). We conclude that the cancellation is happening, and it is happening in each value of isospin separately.

Table 6.2. Sum rule contributions of the lowest pseudoscalar and tensor mesons (last column). The experimental values of meson masses m_M and 2γ decay widths $\Gamma_{\gamma\gamma}$ are from the Particle Data Group [18].

	m_M (MeV)	$\Gamma_{\gamma\gamma}$ (keV)	$\int ds\ \Delta\sigma/s$ (nb)
π^0	134.98	$(7.8 \pm 0.6) \times 10^{-3}$	-195.0 ± 15.0
a_2	1318.3	1.00 ± 0.06	134 ± 8
η	547.85	0.51 ± 0.03	-190.7 ± 11.2
η'	957.66	4.30 ± 0.15	-301.0 ± 10.5
f_2	1275.1	3.03 ± 0.35	448 ± 52
f_2'	1525	0.081 ± 0.009	7 ± 1
Sum η, η'			-492 ± 22
Sum f_2, f_2'			455 ± 53

A more comprehensive and accurate analysis of the sum rule saturation in the meson sector has recently been done by Dai and Pennington [19, 20]. The precision of their analysis is sufficient to distinguish the importance of the 4-pion production channel. The analyses at finite Q^2, using sum rules such as (6.20), allow one to relate the various meson transition form factors, see e.g. [6, 21]. The latter analyses have further been improved by Danilkin and Vanderhaeghen [22], also for the charm mesons [23]. The sum rules and the forward dispersion relations for LbL scattering, in general, have been used to gain insight into the lattice QCD [24, 25], as well as dispersive [26, 27], calculations of the hadronic contributions to muon $g - 2$.

6.8 Composite Higgs

It is interesting to consider how the Higgs boson (H^0) production in $\gamma\gamma$ fusion could affect the helicity-difference sum rules. The Higgs is a neutral scalar particle which can be produced via the mechanism of figure 6.7, where the charged particles in the loop are massive gauge bosons ($W^{\pm}$), leptons, and quarks. This production channel, $\gamma\gamma \to H^0$, contributes only to the helicity-0 cross section, and thus must be accompanied by something else. Something that would mainly affect the helicity-2 cross section and cancel the single-Higgs contribution to the sum rule.

There are no tensor bosons in the Standard Model, and so one needs to consider other mechanisms, e.g. the multiparticle production channels: $\gamma\gamma \to H^0W^+W^-$, $\gamma\gamma \to H^0e^+e^-$, etc. There is little doubt that the sum rule should work within a consistent quantum field theory such as the Standard Model. However, for any physics beyond, and in particular the 'composite Higgs' scenarios, the sum rule will have important implications.

A fascinating example was recently provided by the '750 GeV diphoton excess' signal at the Large Hadron Collider (LHC), hinting at a new state with a significant $\gamma\gamma$ decay width. It was quickly realized [28, 29] that the helicity-difference sum rule would have a preference for the new state to have spin 2.

In addition to the spin, one could provide a rough prediction for its $\gamma\gamma$-width. According to equation (6.32), the exact cancellation will occur when the ratio of the new-particle width over its mass cubed, is $2J + 1 = 5$ times smaller than the corresponding ratio for the Higgs. Assuming the mass of the Higgs and the new state (let's denote it H') is 125 and 750 GeV, respectively, the prediction for the width is

$$\Gamma_{\mathrm{H}'\to\gamma\gamma} \simeq (6^3/5)\Gamma_{\mathrm{H}^0\to\gamma\gamma}. \tag{6.33}$$

So, it would be about 43 times larger than the Higgs decay width.

In the mean time, the 750 GeV signal has disappeared. This exciting scenario is put into a drawer, for a while.

References

[1] Euler H and Kockel B 1935 Ueber die Streuung von Licht an Licht nach der Diracschen Theorie *Naturwiss.* **23** 246

[2] Heisenberg W and Euler H 1936 Folgerungen aus der Diracschen Theorie des Positrons *Z. Phys.* **98** 714–32

[3] Karplus R and Neuman M 1950 Non-linear interactions between electromagnetic fields *Phys. Rev.* **80** 380–5

[4] Karplus R and Neuman M 1951 The scattering of light by light *Phys. Rev.* **83** 776–84

[5] Aaboud M *et al* 2017 Evidence for light-by-light scattering in heavy-ion collisions with the ATLAS detector at the LHC *Nature Phys.* **13** 852–8

[6] Pascalutsa V, Pauk V and Vanderhaeghen M 2012 Light-by-light scattering sum rules constraining meson transition form factors *Phys. Rev.* D **85** 116001

[7] Gerasimov S B and Moulin J 1976 Check of sum rules for photon interaction cross-sections in quantum electrodynamics and mesodynamics *Yad. Fiz.* **23** 142–52
Gerasimov S B and Moulin J 1975 Tests of sum rules for photon total cross sections in quantum electrodynamics and mesodynamics *Nucl. Phys.* B **98** 349

[8] Brodsky S J and Schmidt I 1995 Classical photoabsorption sum rules *Phys. Lett.* B **351** 344–8

[9] Efremov A V and Teryaev O V 1990 Axial anomaly and sum rule for photon spin structure function *Phys. Lett.* B **240** 200

[10] Bass S D 1992 Polarized deep inelastic scattering from photon targets and axial anomalies *Int. J. Mod. Phys.* A **7** 6039–52

[11] Narison S, Shore G M and Veneziano G 1993 A sum rule for the polarized photon structure function $g - 1\gamma$ *Nucl. Phys.* B **391** 69–99

[12] Bass S D, Brodsky S J and Schmidt I 1998 The spin structure of a polarized photon *Phys. Lett.* B **437** 417–24

[13] Pascalutsa V and Vanderhaeghen M 2010 Sum rules for light-by-light scattering *Phys. Rev. Lett.* **105** 201603

[14] Bohm M and Schuster R 1994 Scattering of light by light in the electroweak standard model *Z. Phys.* C **63** 219–25

[15] Pauk V, Pascalutsa V and Vanderhaeghen M 2013 Analytic structure of ϕ^4 theory using light-by-light sum rules *Phys. Lett.* B **725** 504–9

[16] Henley E and Thirring W 1962 *Elementary Quantum Field Theory* (New York: McGraw-Hill)

[17] Pascalutsa V 2016 Surpassing Wigner's causality bound in relativistic scattering with zero-range interaction *PoS CD* **15** 114

[18] Amsler C *et al* 2008 Review of particle physics *Phys. Lett.* B **667** 1

[19] Dai L-Y and Pennington M R 2014 Comprehensive amplitude analysis of $\gamma\gamma \to \pi^+\pi^-$, $\pi^0\pi^0$ and $\bar{K}K$ below 1.5 GeV *Phys. Rev.* D **90** 036004

[20] Dai L-Y and Pennington M 2017 Pascalutsa–Vanderhaeghen light-by-light sum rule from photon–photon collisions *Phys. Rev.* D **95** 056007

[21] Masuda M *et al* 2016 Study of π^0 pair production in single-tag two-photon collisions *Phys. Rev.* D **93** 032003

[22] Danilkin I and Vanderhaeghen M 2017 Light-by-light scattering sum rules in light of new data *Phys. Rev.* D **95** 014019

[23] Danilkin I and Vanderhaeghen M 2017 Light-by-light forward scattering sum rules for charmonium states *Phys. Rev.* D **96** 056003

[24] Green J, Gryniuk O, von Hippel G, Meyer H B and Pascalutsa V 2015 Lattice QCD calculation of hadronic light-by-light scattering *Phys. Rev. Lett.* **115** 222003

[25] Gérardin A, Green J, Gryniuk O, von Hippel G, Meyer H B, Pascalutsa V and Wittig H 2018 Hadronic light-by-light scattering amplitudes from lattice QCD versus dispersive sum rules *Phys. Rev.* D **98** 074501

[26] Colangelo G, Hoferichter M, Procura M and Stoffer P 2017 Rescattering effects in the hadronic-light-by-light contribution to the anomalous magnetic moment of the muon *Phys. Rev. Lett.* **118** 232001

[27] Colangelo G, Hoferichter M, Procura M and Stoffer P 2017 Dispersion relation for hadronic light-by-light scattering: two-pion contributions *JHEP* **04** 161

[28] Roig P and Sanz-Cillero J J 2016 A broad 750 GeV diphoton resonance? Not alone *Phys. Rev.* D **94** 095021

[29] Roig P and Sanz-Cillero J J 2017 Sum-rule constraints on possible diphoton resonances at LHC *EPJ Web Conf.* **137** 10010

IOP Publishing

Causality Rules (Second Edition)

Dispersion theory in non-elementary particle physics

Vladimir Pascalutsa

Chapter 7

Virtual light-by-light scattering

We now consider the elastic scattering of virtual photons.

7.1 Forward scattering amplitudes

In the most general case we consider the forward scattering of virtual photons on virtual photons:

$$\gamma^*(\lambda_1, q_1) + \gamma^*(\lambda_2, q_2) \to \gamma^*(\lambda_1', q_1) + \gamma^*(\lambda_2', q_2), \tag{7.1}$$

where q_1, q_2 are photon four-momenta, and λ_1, λ_2 (λ_1', λ_2') are the helicities of the initial (final) virtual photons, which can take on the values ± 1 (transverse polarizations) and zero (longitudinal). The total helicity in the $\gamma^*\gamma^*$ c.m. system is given by $\Lambda = \lambda_1 - \lambda_2 = \lambda_1' - \lambda_2'$. To define the kinematics, we first introduce the photon virtualities $Q_1^2 = -q_1^2$, $Q_2^2 = -q_2^2$, the Mandelstam invariants, $s = (q_1 + q_2)^2$, $u = (q_1 - q_2)^2$, and the following crossing-symmetric variable:

$$\nu \equiv \frac{1}{4}(s - u) = q_1 \cdot q_2, \tag{7.2}$$

such that $s = 2\nu - Q_1^2 - Q_2^2$, $u = -2\nu - Q_1^2 - Q_2^2$.

The $\gamma^*\gamma^* \to \gamma^*\gamma^*$ forward scattering amplitudes, denoted as $M_{\lambda_1'\lambda_2',\lambda_1\lambda_2}$, are functions of ν, Q_1^2, Q_2^2. Parity invariance (P) and time-reversal invariance (T) imply the following relations among the matrix elements with different helicities:

$$P: \quad M_{\lambda_1'\lambda_2',\lambda_1\lambda_2} = M_{-\lambda_1'-\lambda_2',-\lambda_1-\lambda_2}, \tag{7.3}$$

$$T: \quad M_{\lambda_1'\lambda_2',\lambda_1\lambda_2} = M_{\lambda_1\lambda_2,\lambda_1'\lambda_2'}, \tag{7.4}$$

which leaves out only eight independent amplitudes:

$$M_{++,++},\ M_{+-,+-},\ M_{++,--},\ M_{00,00},\ M_{+0,+0},\ M_{0+,0+},\ M_{++,00},\ M_{0+,-0}. \tag{7.5}$$

We next look at the constraint imposed by crossing symmetry, which requires that the amplitudes for the process (7.1) equal the amplitudes for the process where the photons with e.g. label 2 are crossed:

$$\gamma^*(\lambda_1, q_1) + \gamma^*(-\lambda_2', -q_2) \to \gamma^*(\lambda_1', q_1) + \gamma^*(-\lambda_2, -q_2). \tag{7.6}$$

As under photon crossing $\nu \to -\nu$, one obtains

$$M_{\lambda_1'\lambda_2',\lambda_1\lambda_2}\left(\nu, Q_1^2, Q_2^2\right) = M_{\lambda_1'-\lambda_2,\lambda_1-\lambda_2'}\left(-\nu, Q_1^2, Q_2^2\right), \tag{7.7}$$

it becomes convenient to introduce amplitudes which are either even or odd in ν (at fixed Q_1^2 and Q_2^2). One easily verifies that the following six amplitudes are *even* in ν:

$$(M_{++,++} + M_{+-,+-}), \quad M_{++,--}, \quad M_{00,\,00}, \quad M_{+0,+0}, \quad M_{0+,0+}, \quad (M_{++,00} + M_{0+,-0}), \tag{7.8}$$

whereas the following two amplitudes are *odd* in ν:

$$(M_{++,++} - M_{+-,+-}), \qquad (M_{++,00} - M_{0+,-0}). \tag{7.9}$$

7.1.1 General decomposition of the forward LbL amplitude

References [1, 2] give a decomposition of the absorptive part $W^{\mu'\nu',\,\mu\nu}(q_1, q_2)$ of the forward LbL scattering amplitude. Assume the same decomposition applies to the amplitude itself, $\mathcal{M}^{\mu'\nu',\,\mu\nu}(q_1, q_2)$. This uses the definitions

$$k_1 \equiv \sqrt{\frac{-q_1^2}{X}}\left(q_2 - \frac{q_1 \cdot q_2}{q_1^2}q_1\right), \quad k_2 \equiv \sqrt{\frac{-q_2^2}{X}}\left(q_1 - \frac{q_1 \cdot q_2}{q_2^2}q_2\right), \tag{7.10}$$

$$R^{\alpha\beta} = R^{\beta\alpha} \equiv -g^{\alpha\beta} + \frac{1}{X}\left[\left(q_1 \cdot q_2\right)\left(q_1^\alpha q_2^\beta + q_1^\beta q_2^\alpha\right) - q_1^2 q_2^\alpha q_2^\beta - q_2^2 q_1^\alpha q_1^\beta\right], \tag{7.11}$$

where $X \equiv (q_1 \cdot q_2)^2 - q_1^2 q_2^2$ and the metric is mostly minus. These satisfy $k_i^2 = 1$, $k_i \cdot q_i = 0$, $R^\alpha{}_\beta q_i^\beta = 0$, $R^\alpha{}_\beta R^{\beta\gamma} = -R^{\alpha\gamma}$, and $R^\alpha{}_\alpha = -2$. The decomposition is

$$\begin{aligned}
\mathcal{M}^{\mu'\nu',\,\mu\nu}\left(q_1, q_2\right) = {} & R^{\mu\mu'}R^{\nu\nu'}\mathcal{M}_{\mathrm{TT}} \\
& + \frac{1}{2}[R^{\mu\nu}R^{\mu'\nu'} + R^{\mu\nu'}R^{\mu'\nu} - R^{\mu\mu'}R^{\nu\nu'}]\mathcal{M}^{\tau}_{\mathrm{TT}} \\
& + [R^{\mu\nu}R^{\mu'\nu'} - R^{\mu\nu'}R^{\mu'\nu}]\mathcal{M}^{a}_{\mathrm{TT}} \\
& + R^{\mu\mu'}k_2^\nu k_2^{\nu'}\mathcal{M}_{\mathrm{TL}} + k_1^\mu k_1^{\mu'}R^{\nu\nu'}\mathcal{M}_{\mathrm{LT}} + k_1^\mu k_1^{\mu'}k_2^\nu k_2^{\nu'}\mathcal{M}_{\mathrm{LL}} \\
& - [R^{\mu\nu}k_1^{\mu'}k_2^{\nu'} + R^{\mu\nu'}k_1^{\mu'}k_2^{\nu} + (\mu\nu \leftrightarrow \mu'\nu')]\mathcal{M}^{\tau}_{\mathrm{TL}} \\
& - [R^{\mu\nu}k_1^{\mu'}k_2^{\nu'} - R^{\mu\nu'}k_1^{\mu'}k_2^{\nu} + (\mu\nu \leftrightarrow \mu'\nu')]\mathcal{M}^{a}_{\mathrm{TL}},
\end{aligned} \tag{7.12}$$

where the eight amplitudes are functions of q_1^2, q_2^2, and $\nu \equiv q_2 \cdot q_2$. The prefactors are orthogonal, and we obtain

$$
\begin{aligned}
\mathcal{M}_{\rm TT} &= \frac{1}{4} R^{\mu\mu'} R^{\nu\nu'} \mathcal{M}_{\mu'\nu',\mu\nu} \\
\mathcal{M}^{\tau}_{\rm TT} &= \frac{1}{4} [R^{\mu\nu} R^{\mu'\nu'} + R^{\mu\nu'} R^{\mu'\nu} - R^{\mu\mu'} R^{\nu\nu'}] \mathcal{M}_{\mu'\nu',\mu\nu} \\
\mathcal{M}^{a}_{\rm TT} &= \frac{1}{4} [R^{\mu\nu} R^{\mu'\nu'} - R^{\mu\nu'} R^{\mu'\nu}] \mathcal{M}_{\mu'\nu',\mu\nu} \\
\mathcal{M}_{\rm TL} &= \frac{1}{2} R^{\mu\mu'} k_2^{\nu} k_2^{\nu'} \mathcal{M}_{\mu'\nu',\mu\nu} \\
\mathcal{M}_{\rm LT} &= \frac{1}{2} k_1^{\mu} k_1^{\mu'} R^{\nu\nu'} \mathcal{M}_{\mu'\nu',\mu\nu} \\
\mathcal{M}_{\rm LL} &= k_1^{\mu} k_1^{\mu'} k_2^{\nu} k_2^{\nu'} \mathcal{M}_{\mu'\nu',\mu\nu} \\
\mathcal{M}^{\tau}_{\rm TL} &= -\frac{1}{8} [R^{\mu\nu} k_1^{\mu'} k_2^{\nu'} + R^{\mu\nu'} k_1^{\mu'} k_2^{\nu} + (\mu\nu \leftrightarrow \mu'\nu')] \mathcal{M}_{\mu'\nu',\mu\nu} \\
\mathcal{M}^{a}_{\rm TL} &= -\frac{1}{8} [R^{\mu\nu} k_1^{\mu'} k_2^{\nu'} - R^{\mu\nu'} k_1^{\mu'} k_2^{\nu} + (\mu\nu \leftrightarrow \mu'\nu')] \mathcal{M}_{\mu'\nu',\mu\nu}.
\end{aligned}
\tag{7.13}
$$

Crossing symmetry implies $\mathcal{M}^{\mu'\nu',\mu\nu}(q_1, q_2) = \mathcal{M}^{\mu\nu',\mu'\nu}(-q_1, q_2)$. Under $q_1 \to -q_1$, we find that $k_2 \to -k_2$, whereas k_1 and $R^{\alpha\beta}$ are unchanged. We thus find that $\mathcal{M}^{\tau}_{\rm TL}$ is odd in ν and the other amplitudes are even[1]. For example (here we show explicitly the dependence on q_1 and q_2),

$$
\begin{aligned}
\mathcal{M}^{\tau}_{\rm TL}(q_1^2, q_2^2, \nu) &= -\frac{1}{8}\Big[R^{\mu\nu}(q_1, q_2) k_1^{\mu'}(q_1, q_2) k_2^{\nu'}(q_1, q_2) \\
&\quad + R^{\mu\nu'}(q_1, q_2) k_1^{\mu'}(q_1, q_2) k_2^{\nu}(q_1, q_2) \\
&\quad + (\mu\nu \leftrightarrow \mu'\nu')\Big] \mathcal{M}_{\mu'\nu',\mu\nu}(q_1, q_2) \\
(q_1 \to -q_1) \qquad &= +\frac{1}{8}\Big[R^{\mu\nu}(-q_1, q_2) k_1^{\mu'}(-q_1, q_2) k_2^{\nu'}(-q_1, q_2) \\
&\quad + R^{\mu\nu'}(-q_1, q_2) k_1^{\mu'}(-q_1, q_2) k_2^{\nu}(-q_1, q_2) \\
&\quad + (\mu\nu \leftrightarrow \mu'\nu')\Big] \mathcal{M}_{\mu\nu',\mu'\nu}(-q_1, q_2) \\
(\text{rearrangeterms}) \qquad &= +\frac{1}{8}\Big[R^{\mu'\nu}(-q_1, q_2) k_1^{\mu}(-q_1, q_2) k_2^{\nu'}(-q_1, q_2) \\
&\quad + R^{\mu'\nu'}(-q_1, q_2) k_1^{\mu}(-q_1, q_2) k_2^{\nu}(-q_1, q_2) \\
&\quad + (\mu\nu \leftrightarrow \mu'\nu')\Big] \mathcal{M}_{\mu\nu',\mu'\nu}(-q_1, q_2) \\
(\text{relabel}\mu \leftrightarrow \mu') &= -\mathcal{M}^{\tau}_{\rm TL}(q_1^2, q_2^2, -\nu).
\end{aligned}
\tag{7.14}
$$

[1] According to [2, 3], $\mathcal{M}^{a}_{\rm TT}$ and $\mathcal{M}^{a}_{\rm TL}$ are odd, whereas [1] says the opposite.

7.1.2 Unitarity

The optical theorem relates the absorptive parts of the LbL scattering amplitudes to $\gamma^*\gamma^*\to X$ cross sections:

$$\begin{aligned}
W_{\mathrm{TT}} &= 2\sqrt{X}\sigma_{\mathrm{TT}} = \sqrt{X}\left(\sigma_0 + \sigma_2\right) = \sqrt{X}\left(\sigma_{\parallel} + \sigma_{\perp}\right) \\
W_{\mathrm{TT}}^{\tau} &= 2\sqrt{X}\tau_{\mathrm{TT}} = 2\sqrt{X}\left(\sigma_{\parallel} - \sigma_{\perp}\right) \\
W_{\mathrm{TT}}^{a} &= 2\sqrt{X}\tau_{\mathrm{TT}}^{a} = \sqrt{X}(\sigma_0 - \sigma_2) \\
W_{\mathrm{TL}} &= 2\sqrt{X}\sigma_{\mathrm{TL}} \qquad W_{\mathrm{LT}} = 2\sqrt{X}\sigma_{\mathrm{LT}} \qquad W_{\mathrm{LL}} = 2\sqrt{X}\sigma_{\mathrm{LL}} \\
W_{\mathrm{TL}}^{\tau} &= 2\sqrt{X}\tau_{\mathrm{TL}} \qquad W_{\mathrm{TL}}^{a} = 2\sqrt{X}\tau_{\mathrm{TL}}^{a},
\end{aligned} \tag{7.15}$$

where each W is the imaginary part of the corresponding amplitude $\mathcal{M}$. All of the cross sections denoted with σ are positive definite.

7.1.3 Dispersion relations

As discussed in chapter 2, relativistic causality at the quantum level (microcausality) translates into exact statements about analytic properties of the scattering amplitude in the complex energy plane. In this case, it is the statement of analyticity of the forward $\gamma^*\gamma^*$ scattering amplitude in the entire ν plane, except for the real axis where the branch cuts associated with particle production are located. Assuming that the threshold for particle production is $\nu_0 > 0$, one can write down the usual dispersion relations, in which the amplitude is given by integrals over the non-analyticities, which in this case are branch cuts extending from $\pm\nu_0$ to $\pm\infty$. Finally, for amplitudes that are even or odd in ν we can write (for any fixed values of Q_1^2, $Q_2^2 > 0$):

$$f_{\mathrm{even}}(\nu) = \frac{2}{\pi}\int_{\nu_0}^{\infty} d\nu' \frac{\nu'}{\nu'^2 - \nu^2 - i0^+} \operatorname{Im} f_{\mathrm{even}}(\nu'), \tag{7.16a}$$

$$f_{\mathrm{odd}}(\nu) = \frac{2\nu}{\pi}\int_{\nu_0}^{\infty} d\nu' \frac{1}{\nu'^2 - \nu^2 - i0^+} \operatorname{Im} f_{\mathrm{odd}}(\nu'), \tag{7.16b}$$

where 0^+ is an infinitesimal positive number.

These dispersion relations are derived with the provision that the integrals converge. If they do not, subtractions must be made; e.g. the once-subtracted dispersion relation for the even amplitudes reads:

$$f_{\mathrm{even}}(\nu) = f_{\mathrm{even}}(0) + \frac{2\nu^2}{\pi}\int_{\nu_0}^{\infty} d\nu' \frac{1}{\nu'(\nu'^2 - \nu^2 - i0^+)} \operatorname{Im} f_{\mathrm{even}}(\nu'). \tag{7.17}$$

We therefore need the information about these amplitudes at zero energy ν. Anticipating the discussion of the low-energy expansion of the LbL scattering, we can state that at $\nu = 0$ these amplitudes vanish when one of the photons is real (see equation (7.21)). Using equation (7.15) to substitute the cross sections in place of the absorptive parts, we obtain the following sum rules for the case of one real and one virtual photon (when the virtual photon flux factor becomes $X = \nu^2$):

$$M_{++,++}(\nu) + M_{+-,+-}(\nu) = \frac{4\nu^2}{\pi} \int_{\nu_0}^{\infty} d\nu' \, \frac{\sigma_{\parallel}(\nu') + \sigma_{\perp}(\nu')}{\nu'^2 - \nu^2 - i0^+}, \tag{7.18a}$$

$$M_{++,--}(\nu) = \frac{4\nu^2}{\pi} \int_{\nu_0}^{\infty} d\nu' \, \frac{\sigma_{\parallel}(\nu') - \sigma_{\perp}(\nu')}{\nu'^2 - \nu^2 - i0^+}, \tag{7.18b}$$

$$M_{0+,0+}(\nu) = \frac{4\nu^2}{\pi} \int_{\nu_0}^{\infty} d\nu' \, \frac{\sigma_{\mathrm{LT}}(\nu')}{\nu'^2 - \nu^2 - i0^+}, \tag{7.18c}$$

$$M_{+0,+0}(\nu) = \frac{4\nu^2}{\pi} \int_{\nu_0}^{\infty} d\nu' \, \frac{\sigma_{\mathrm{TL}}(\nu')}{\nu'^2 - \nu^2 - i0^+}. \tag{7.18d}$$

We cannot write such a subtracted sum rule for $M_{00,\,00}$, since it trivially vanishes when one of the photons is real. Instead, we consider an unsubtracted dispersion relation:

$$M_{00,\,00}(\nu) = \frac{4}{\pi} \int_{\nu_0}^{\infty} d\nu' \, \frac{\nu' \sqrt{X'} \, \sigma_{\mathrm{LL}}(\nu')}{\nu'^2 - \nu^2 - i0^+}, \tag{7.18e}$$

with $X' = \nu'^2 - Q_1^2 Q_2^2$. It is not difficult to check that, at least in perturbative QED calculations (see section 7.2), this integral converges which validates this unsubtracted relation renormalizable, perturbative field theory. This is usually all we need to be confident that the unsubtracted relation works.

For all the remaining amplitudes the asymptotic behavior justifies the use of unsubtracted dispersion relations which, upon inserting the unitarity conditions (7.15), lead to the following dispersion relations, valid for arbitrary space-like virtualities:

$$M_{++,++}(\nu) - M_{+-,+-}(\nu) = \frac{4\nu}{\pi} \int_{\nu_0}^{\infty} d\nu' \, \frac{\sqrt{X'} \, [\sigma_0(\nu') - \sigma_2(\nu')]}{\nu'^2 - \nu^2 - i0^+}, \tag{7.18f}$$

$$M_{++,00}(\nu) - M_{0+,-0}(\nu) = \frac{8\nu}{\pi} \int_{\nu_0}^{\infty} d\nu' \, \frac{\sqrt{X'} \, \tau^a_{\mathrm{TL}}(\nu')}{\nu'^2 - \nu^2 - i0^+}, \tag{7.18g}$$

$$M_{++,00}(\nu) + M_{0+,-0}(\nu) = \frac{8}{\pi} \int_{\nu_0}^{\infty} d\nu' \, \frac{\nu' \sqrt{X'} \, \tau_{\mathrm{TL}}(\nu')}{\nu'^2 - \nu^2 - i0^+}, \tag{7.18h}$$

even if the dependence on virtualities Q_1^2, Q_2^2 is not written out explicitly.

The above sum rules, relating all the forward $\gamma^*\gamma^*$ elastic scattering amplitudes to the energy integrals of the $\gamma^*\gamma^*$ fusion cross sections, should hold for any space-like photon virtualities in the unsubtracted cases, and for one of the virtualities equal to zero in the subtracted cases. In the following we examine the low-energy expansion of these sum rules.

7.1.4 Low-energy expansion via an effective Lagrangian

To obtain more specific relations from the sum rules established in equation (7.18a), we parametrize the low-energy (small ν) behavior of the $\gamma^*\gamma^*\to\gamma^*\gamma^*$ forward scattering amplitudes M. At lowest order in the energy, the self-interactions of the electromagnetic field are described by an effective Lagrangian (of fourth order in the photon energy and/or momentum, and fourth order in the electromagnetic field):

$$\mathcal{L}^{(8)} = c_1\left(F_{\mu\nu}F^{\mu\nu}\right)^2 + c_2\left(F_{\mu\nu}\tilde{F}^{\mu\nu}\right)^2, \tag{7.19}$$

where $F_{\mu\nu} = \partial_\mu A_\nu - \partial_\nu A_\mu$, $\tilde{F}^{\mu\nu} = \varepsilon^{\mu\nu\alpha\beta}\partial_\alpha A_\beta$, and where c_1, c_2 are two low-energy constants (LECs) which contain the structure dependent information.

At the next order in energy, one considers the terms involving two derivatives on the field tensors, corresponding with the sixth order in the photon energy and/or momentum. Writing down all such dimension-ten operators and reducing their number using the antisymmetry of the field tensors, the Bianchi identities, as well as adding or removing total derivative terms, we find that there are six independent terms at that order, which we choose as

$$\begin{aligned}\mathcal{L}^{(10)} = {} & c_3\left(\partial_\alpha F_{\mu\nu}\right)\left(\partial^\alpha F^{\lambda\nu}\right)F_{\lambda\rho}F^{\mu\rho} + c_4\left(\partial_\alpha F_{\mu\nu}\right)\left(\partial^\alpha F^{\mu\nu}\right)F_{\lambda\rho}F^{\lambda\rho} \\ & + c_5\left(\partial^\alpha F_{\alpha\nu}\right)\left(\partial_\beta F^{\beta\nu}\right)F_{\lambda\rho}F^{\lambda\rho} + c_6\left(\partial_\alpha\partial^\alpha F_{\mu\nu}\right)F^{\lambda\nu}F_{\lambda\rho}F^{\mu\rho} \\ & + c_7\left(\partial_\alpha\partial^\alpha F_{\mu\nu}\right)F^{\mu\nu}F_{\lambda\rho}F^{\lambda\rho} + c_8\left(\partial^\alpha F_{\alpha\mu}\right)\left(\partial_\beta F^{\beta\lambda}\right)F_{\rho\lambda}F^{\rho\mu},\end{aligned} \tag{7.20}$$

where $c_3, \ldots, c_8$ are the new LECs arising at this order. Only c_3 and c_4 appear in the case of real photons.

We can now specify the low-energy limit of the light-by-light scattering amplitudes in terms of the LECs describing the low-energy self-interactions of the electromagnetic field:

$$\begin{aligned}M_{++,++} + M_{+-,+-} = {} & Q_1^2Q_2^2\left[64\left(c_1 - c_2\right) + 4\left(Q_1^2 + Q_2^2\right)\left(-c_3 - 8c_4 - 4c_5 + 8c_7 - c_8\right) + \mathcal{O}(Q^4)\right] \\ & + 8\nu^2\left[8\left(c_1 + c_2\right) + \left(Q_1^2 + Q_2^2\right)\left(-c_3 + 3c_6 + 4c_7\right) + \mathcal{O}(Q^4)\right] + \mathcal{O}(\nu^4),\end{aligned} \tag{7.21a}$$

$$\begin{aligned}M_{++,--} = {} & Q_1^2Q_2^2[64c_2 + 4(Q_1^2 + Q_2^2)(-c_3 + 2c_6 - c_8) + \mathcal{O}(Q^4)] \\ & + 8\nu^2[8(c_1 - c_2) + (Q_1^2 + Q_2^2)(c_6 + 4c_7) + \mathcal{O}(Q^4)] + \mathcal{O}(\nu^4),\end{aligned} \tag{7.21b}$$

$$\begin{aligned}M_{0+,0+} = {} & Q_1^2Q_2^2\left[-32c_1 + 4Q_1^2c_8 + 4\left(Q_1^2 + Q_2^2\right)\left(c_3 + 4c_4 + 2c_5 - 2c_6 - 4c_7\right) + \mathcal{O}(Q^4)\right] \\ & + \nu^2\left[-4Q_1^2c_8 + \mathcal{O}(Q^4)\right] + \mathcal{O}(\nu^4),\end{aligned} \tag{7.21c}$$

$$\begin{aligned}M_{+0,+0} = {} & Q_1^2Q_2^2\left[-32c_1 + 4Q_2^2c_8 + 4\left(Q_1^2 + Q_2^2\right)\left(c_3 + 4c_4 + 2c_5 - 2c_6 - 4c_7\right) + \mathcal{O}(Q^4)\right] \\ & + \nu^2\left[-4Q_2^2c_8 + \mathcal{O}(Q^4)\right] + \mathcal{O}(\nu^4),\end{aligned} \tag{7.21d}$$

$$M_{00,\,00} = Q_1^2 Q_2^2\Big[96c_1 + 4\big(Q_1^2 + Q_2^2\big)\big(-2c_3 - 4c_4 - 2c_5 + 6c_6 + 12c_7 - c_8\big) + \mathcal{O}(Q^4)\Big] + \mathcal{O}(\nu^2), \tag{7.21e}$$

$$M_{++,++} - M_{+-,+-} = 8\nu Q_1^2 Q_2^2[-c_3 - 4c_5 + c_8 + \mathcal{O}(Q^2)] + \nu^3[-64c_4 + \mathcal{O}(Q^2)] + \mathcal{O}(\nu^5), \tag{7.21f}$$

$$M_{++,00} - M_{0+,-0} = \nu Q_1 Q_2\Big[-64c_1 + \big(Q_1^2 + Q_2^2\big)\big(4c_3 - 16c_6 - 32c_7 + 4c_8\big) + \mathcal{O}(Q^4)\Big] + \mathcal{O}(\nu^3), \tag{7.21g}$$

$$M_{++,00} + M_{0+,-0} = Q_1^3 Q_2^3[4c_5 - 12c_8 + \mathcal{O}(Q^2)] + 4\nu^2 Q_1 Q_2[2c_3 + 16c_4 + 4c_5 + c_8 + \mathcal{O}(Q^2)] + \mathcal{O}(\nu^4). \tag{7.21h}$$

These expressions can be treated as a simultaneous expansion in ν and the virtualities Q_i^2 of the lhs of the sum rules equation (7.18). Concerning the Q dependence, it is important that the leading in ν term, in any of the amplitudes, is proportional to $Q_1 Q_2$ and hence vanishes for at least one real photon. The latter statement is valid for any values of virtualities, not just when they are small. For example, let us show for the amplitude $(M_{++,++} - M_{+-,+-})$ its leading term in ν is proportional to the combination $Q_1^2 Q_2^2$, to all orders in Q_1 and Q_2.

Since all photons are transversely polarized the only non-vanishing structures involving polarization vectors of photons $\varepsilon(\lambda_i)$ are their mutual scalar products $\varepsilon(\lambda_i)\cdot\varepsilon(\lambda_j)$. Due to gauge invariance, the electromagnetic fields enter the Lagrangian through the field tensor $F_{\mu\nu}$, which contributes to the amplitude as $q_\mu\varepsilon_\nu - q_\nu\varepsilon_\mu$. Thus an arbitrary term in the effective Lagrangian contributes to $(M_{++,++} - M_{+-,+-})$ as

$$M_{++,++} - M_{+-,+-} \sim q_1^\mu q_2^\nu q_1^\lambda q_2^\rho T_{\mu\nu\lambda\rho}, \tag{7.22}$$

where the tensor $T_{\mu\nu\lambda\rho}$ is constructed from four-vectors q_i and the metric tensor. Since this amplitude is odd with respect to ν, it is required to be proportional to at least ν^1. Assuming that one factor ν comes from contraction of two of the q in equation (7.22), we are left with $q_1^\mu q_2^\nu$. Now, if we suppose that q_1 is contracted with q_2 we obtain an extra power of ν, and such an amplitude vanishes when taking the limit $\nu \to 0$. Thus, both q_1 and q_2 must be contracted with another q_1 and q_2 respectively, giving a global factor $Q_1^2 Q_2^2$.

We are now in position to examine the sum rules in equation (7.18) order by order in ν. For this we expand the rhs of equation (7.18) using

$$1/(\nu'^2 - \nu^2) = 1/\nu'^2 + \nu^2/\nu'^4 + \mathcal{O}(\nu^4)$$

. As the result, equations (7.18f), (7.18g), and (7.18h) yield the following super-convergence relations, valid for at least one real photon (e.g. $Q_1 \geqslant 0$, $Q_2^2 = 0$):

$$0 = \int_{s_0}^{\infty} ds \frac{1}{\left(s + Q_1^2\right)}\, \tau_{\mathrm{TT}}^a(s,\, Q_1^2,\, 0), \tag{7.23a}$$

$$0 = \int_{s_0}^{\infty} ds \frac{1}{\left(s + Q_1^2\right)^2} \left[\sigma_{\|} + \sigma_{\mathrm{LT}} + \frac{\left(s + Q_1^2\right)}{Q_1 Q_2} \tau_{\mathrm{TL}}^a \right]_{Q_2^2=0}, \tag{7.23b}$$

$$0 = \int_{s_0}^{\infty} ds \left[\frac{\tau_{\mathrm{TL}}\left(s, Q_1^2, Q_2^2\right)}{Q_1 Q_2} \right]_{Q_2^2=0}, \tag{7.23c}$$

and the following set of sum rules for the LECs of the dimension-8 effective Lagrangian, valid when both photons are quasi-real:

$$c_1 = \frac{1}{8\pi} \int_{s_0}^{\infty} \frac{ds}{s^2} \sigma_{\|}\left(s, 0, 0\right), \tag{7.24a}$$

$$= -\frac{1}{8\pi} \int_{s_0}^{\infty} \frac{ds}{s} \left[\frac{\tau_{\mathrm{TL}}^a\left(s, Q_1^2, Q_2^2\right)}{Q_1 Q_2} \right]_{Q_1^2=Q_2^2=0}, \tag{7.24b}$$

$$= \frac{1}{8\pi} \int_{s_0}^{\infty} ds \left[\frac{\sigma_{\mathrm{LL}}\left(s, Q_1^2, Q_2^2\right)}{Q_1^2 Q_2^2} \right]_{Q_1^2=Q_2^2=0}, \tag{7.24c}$$

$$c_2 = \frac{1}{8\pi} \int_{s_0}^{\infty} \frac{ds}{s^2} \sigma_{\perp}(s, 0, 0), \tag{7.24d}$$

where $s_0 = 2\nu_0 - Q_1^2 - Q_2^2$.

There are as well the sum rules for the LECs of the dimension-10 Lagrangian, most notably

$$c_4 = -\frac{1}{4\pi} \int_{s_0}^{\infty} \frac{ds}{s^3} \tau_{\mathrm{TT}}^a(s, 0, 0), \tag{7.25}$$

but currently they are of far lesser importance and we do not write them out here explicitly.

Let us note that the relation (7.23a), obtained by combining equations (7.18f) and (7.21f), is the generalized GDH sum rule for the photon target (6.20), which can be cast into the superconvergence relation for the photonspin-structure function g_1^γ: $\int_0^1 dx\, g_1^\gamma(x, Q^2) = 0$, see section 8.4 for more detail.

In the following section we verify these sum rules in QED at leading order in the fine-structure constant α.

7.2 Sum rules in perturbation theory

We will subsequently discuss the tree-level pair production in scalar QED (e.g. $\gamma^*\gamma^* \to \pi^+\pi^-$, see figure 6.2) and in spinor QED ($\gamma^*\gamma^* \to e^+e^-$, see figure 3.1).

7.2.1 Scalar QED

The response functions for the case of leading-order scalar QED, for one of the photons real or quasi-real ($Q_2^2 \to 0$), are given by

$$
\begin{aligned}
[\sigma_\parallel + \sigma_\perp]_{Q_2^2=0} &= \frac{\alpha^2 4\pi s^2}{\left(s+Q_1^2\right)^3}\left\{\sqrt{1-\frac{4m^2}{s}}\left(1+\frac{4m^2}{s}+\frac{Q_1^4}{s^2}\right)-\frac{8m^2}{s}\left(1-\frac{2m^2}{s}-\frac{Q_1^2}{s}\right)L\right\},\\
[\sigma_\parallel - \sigma_\perp]_{Q_2^2=0} &= \frac{\alpha^2 4\pi s^2}{\left(s+Q_1^2\right)^3}\left\{\sqrt{1-\frac{4m^2}{s}}\left(\frac{2m^2}{s}+\frac{Q_1^4}{s^2}\right)+\frac{8m^2}{s}\left(\frac{m^2}{s}+\frac{Q_1^2}{s}\right)L\right\},\\
[\sigma_0 - \sigma_2]_{Q_2^2=0} &= \frac{\alpha^2 4\pi s}{\left(s+Q_1^2\right)^2}\left\{-\sqrt{1-\frac{4m^2}{s}}\left(1-\frac{Q_1^2}{s}\right)+\frac{8m^2}{s}L\right\},\\
\left[\frac{1}{Q_1^2 Q_2^2}\sigma_{LL}\right]_{Q_2^2=0} &= \frac{\alpha^2 8\pi s^2}{\left(s+Q_1^2\right)^5}\left\{\sqrt{1-\frac{4m^2}{s}}\left(8+\frac{s}{4m^2}\left(1-\frac{Q_1^2}{s}\right)^2\right)-\left(7+\frac{Q_1^2}{s}\right)\left(1-\frac{Q_1^2}{s}\right)L\right\},\\
\left[\frac{1}{Q_1^2}\sigma_{LT}\right]_{Q_2^2=0} &= \frac{\alpha^2 4\pi s}{\left(s+Q_1^2\right)^3}\left\{-3\sqrt{1-\frac{4m^2}{s}}+2\left(1+\frac{2m^2}{s}\right)L\right\},\\
\left[\frac{1}{Q_1 Q_2}\tau_{TL}\right]_{Q_2^2=0} &= \frac{\alpha^2 4\pi s}{\left(s+Q_1^2\right)^3}\left\{-\sqrt{1-\frac{4m^2}{s}}+\left(1-\frac{Q_1^2}{s}-\frac{4m^2}{s}\right)L\right\},\\
\left[\frac{1}{Q_1 Q_2}\tau_{TL}^{a}\right]_{Q_2^2=0} &= \frac{\alpha^2 8\pi s^2}{\left(s+Q_1^2\right)^4}\left\{\sqrt{1-\frac{4m^2}{s}}\left(1-\frac{2Q_1^2}{s}\right)-\left[\frac{1}{2}\left(1-\frac{Q_1^2}{s}\right)^2+\frac{4m^2}{s}\right]L\right\},
\end{aligned}
\tag{7.26}
$$

with $\sqrt{s} > 2m$ the total invariant energy and

$$
L = \log\left[\frac{\sqrt{s}}{2m}\left(1+\sqrt{1-\frac{4m^2}{s}}\right)\right] = \cosh^{-1}\frac{\sqrt{s}}{2m}. \tag{7.27}
$$

It is now easy to verify the three superconvergent relations in equation (7.23) for arbitrary space-like virtuality ($Q_1^2 \geqslant 0$) of the virtual photon. All of the integrands exhibit the same pattern of an exact cancellation between low and higher energies, which we have seen in figure 6.3. Also, the four sum rules in equation (7.24) work exactly reproducing the well-known results:

$$
c_1 = \frac{\alpha^2}{m^4}\frac{7}{1440}, \qquad c_2 = \frac{\alpha^2}{m^4}\frac{1}{1440}. \tag{7.28}
$$

7.2.2 Spinor QED

The response functions for the tree-level process $\gamma^*\gamma \to e^+e^-$, where one of the photons is real or quasi-real ($Q_2^2 \to 0$), are given by

$$[\sigma_\parallel + \sigma_\perp]_{Q_2^2=0} = \frac{\alpha^2 8\pi s^2}{\left(s+Q_1^2\right)^3}\left\{\sqrt{1-\frac{4m^2}{s}}\left[-\left(1-\frac{Q_1^2}{s}\right)^2-\frac{4m^2}{s}\right]+2\left(1+\frac{4m^2}{s}-\frac{8m^4}{s^2}+\frac{Q_1^4}{s^2}\right)L\right\},$$

$$[\sigma_\parallel - \sigma_\perp]_{Q_2^2=0} = -\frac{\alpha^2 8\pi s^2}{\left(s+Q_1^2\right)^3}\left\{\sqrt{1-\frac{4m^2}{s}}\left(\frac{2m^2}{s}+\frac{Q_1^4}{s^2}\right)+\frac{8m^2}{s}\left(\frac{m^2}{s}+\frac{Q_1^2}{s}\right)L\right\},$$

$$[\sigma_0 - \sigma_2]_{Q_2^2=0} = \frac{\alpha^2 8\pi s}{\left(s+Q_1^2\right)^2}\left\{\sqrt{1-\frac{4m^2}{s}}\left(3-\frac{Q_1^2}{s}\right)-2\left(1-\frac{Q_1^2}{s}\right)L\right\},$$

$$\left[\frac{1}{Q_1^2 Q_2^2}\sigma_{\mathrm{LL}}\right]_{Q_2^2=0} = \frac{\alpha^2 128\pi s^2}{\left(s+Q_1^2\right)^5}\left\{-\sqrt{1-\frac{4m^2}{s}}+L\right\}, \tag{7.29}$$

$$\left[\frac{1}{Q_1^2}\sigma_{\mathrm{LT}}\right]_{Q_2^2=0} = \frac{\alpha^2 16\pi s}{\left(s+Q_1^2\right)^3}\left\{\sqrt{1-\frac{4m^2}{s}}-\frac{4m^2}{s}L\right\},$$

$$\left[\frac{1}{Q_1 Q_2}\tau_{\mathrm{TL}}\right]_{Q_2^2=0} = 0,$$

$$\left[\frac{1}{Q_1 Q_2}\tau^a_{\mathrm{TL}}\right]_{Q_2^2=0} = \frac{\alpha^2 16\pi s^2}{\left(s+Q_1^2\right)^4}\left\{-\sqrt{1-\frac{4m^2}{s}}\left(1-\frac{2Q_1^2}{s}\right)+\left(-\frac{2Q_1^2}{s}+\frac{4m^2}{s}\right)L\right\},$$

and again, we have verified the three superconvergent relations in equation (7.23) for arbitrary space-like virtuality ($Q_1^2 \geqslant 0$) of the virtual photon. We have verified the sum rules (7.24), reproducing

$$c_1 = \frac{\alpha^2}{m^4}\frac{7}{1440}, \qquad c_2 = \frac{\alpha^2}{m^4}\frac{1}{1440}, \tag{7.30}$$

in the limit of $Q_1^2 \to 0$. In the latter case we have also been able to verify the novel sum rule of equation (7.25), which yields

$$c_4 = -\frac{\alpha^2}{m^6}\frac{1}{315}, \tag{7.31}$$

in agreement with the result of [4], obtained in a direct calculation of the low-energy photon–photon scattering amplitude.

More generally, one can perform a QED verification of the entire dispersion relations, rather than just of the sum rules which follow from their low-energy expansion. This is straightforward, but technically much more involved. Some of these checks can be found in [5].

References

[1] Budnev V, Chernyak V and Ginzburg I 1971 Kinematics of gamma gamma scattering *Nucl. Phys.* B **34** 470–6

[2] Budnev V, Ginzburg I, Meledin G and Serbo V 1975 The two photon particle production mechanism. Physical problems. Applications. Equivalent photon approximation *Phys. Rep.* **15** 181–281

[3] Pascalutsa V, Pauk V and Vanderhaeghen M 2012 Light-by-light scattering sum rules constraining meson transition form factors *Phys. Rev.* D **85** 116001

[4] Dicus D A, Kao C and Repko W W 1998 Effective Lagrangians and low-energy photon–photon scattering *Phys. Rev.* D **57** 2443–7

[5] Biloshytskyi V, Chao E-H, Gérardin A, Green J R, Hagelstein F, Meyer H B, Parrino J and Pascalutsa V 2023 Forward light-by-light scattering and electromagnetic correction to hadronic vacuum polarization *JHEP* **03** 194

Chapter 8

Compton-scattering sum rules for vector bosons

In this chapter we enter the relatively unexplored area of sum rules for a spin-1 target. This subject is interesting in connection with the massive gauge bosons of the Standard Model ($W^{\pm}$, Z^0), but the formalism is just as well applicable to the (virtual) massless gauge bosons, such as photons and gluons. The sum rules for the photon structure functions can simply be obtained in the limit of vanishing electromagnetic moments; we will come to that in the last section. We shall start, however, by considering a charged massive spin-1 particle having a magnetic and quadrupole moments, and more generally having arbitrary electromagnetic distributions (or, form factors). The practical applications are therefore not limited to gauge bosons; the resulting sum rules will apply, for example, to the deuteron or vector mesons (ρ, ω, etc) as well. We shall see that consistent electrodynamics of spin-1 fields is only realized within the Yang–Mills theory, which sets the classical values of the electromagnetic moments to the, so-called, natural values. Some extra attention will be paid to the role of the quadrupole moment, which is of course a novel feature of the spin-1 case, as compared with spin-1/2. It, for instance, affects the sum sum rule for the forward spin polarizability. We shall conclude with a look at the spin-1 structure functions and the sum rules for the photon structure functions in particular.

8.1 Electromagnetic moments: natural values

A massive particle with spin S has, in general, $2S + 1$ intrinsic electromagnetic moments. In the case of $S = 1$, these are: the electric charge e, the magnetic dipole moment μ, and the electric quadrupole moment $\mathcal{Q}$. It is interesting that for an elementary (structureless) charged particle the electromagnetic moments are set to the so-called *natural values* [1–4]. For example, in the spin-1 case, they are given by

$$\mu_{\text{nat.}} = \frac{e}{M}, \qquad \mathcal{Q}_{\text{nat.}} = -\frac{e}{M^2}, \tag{8.1}$$

where M is the mass. The natural value of the magnetic moment corresponds with the gyromagnetic ratio g (i.e. the ratio of the magnetic moment value, in units of the magneton: $e/2M$, to the value of the spin) equal to 2. In fact, it is the natural value for any spin:

$$g_{\text{nat.}} = 2 = \frac{\mu_{\text{nat.}}}{(e/2M)S}. \tag{8.2}$$

As shown in [3], the natural values arise from the point-like transverse charge density. In simpler terms, this means the charge can only be distributed along the direction of motion. The deviations of the electromagnetic moments from their natural values are called the anomalous moments; we hereby introduce them in the dimensionless fashion:

$$\mu = \mu_{\text{nat.}} + 2\varkappa\frac{e}{2M}, \qquad \mathcal{Q} = \mathcal{Q}_{\text{nat.}} + 2\varsigma\frac{e}{M^2}, \tag{8.3}$$

where $\varkappa$ and ς are, respectively, the magnetic and quadrupole anomalous moments. The normalization is chosen such that $\varkappa = (g - 2)/2$, just as for the spin-1/2 particle.

Let us consider a field-theoretic description of a massive spin-1 particle with the above-mentioned intrinsic moments. In relativistic theory, the particle is described by a bosonic vector field $W^\mu(x)$. This is very similar to the electromagnetic field $A^\mu(x)$, except for a charged particle the field is complex. The free Lagrangians for the electromagnetic and vector-boson fields are similar too[1]:

$$\mathcal{L}_{\text{em}} = -\frac{1}{4}F_{\mu\nu}F^{\mu\nu}, \qquad F_{\mu\nu} = \partial_\mu A_\nu - \partial_\nu A_\mu, \tag{8.4a}$$

$$\mathcal{L}_{\text{Proca}} = -\frac{1}{2}W^*_{\mu\nu}W^{\mu\nu} - M^2 W^*_\mu W^\mu, \qquad W_{\mu\nu} = \partial_\mu W_\nu - \partial_\nu W_\mu. \tag{8.4b}$$

The only real difference is the mass term; the factor of 2 difference arises because of the convention to write the complex field in terms of two real vectors W_1^μ and W_2^μ:

$$W^\mu = \frac{1}{\sqrt{2}}(W_1^\mu - iW_2^\mu). \tag{8.5}$$

Substituting this into the above Lagrangian, one can see that the normalization factors for the real fields are all the same.

In the field-theoretic language, the electromagnetic moments arise from the cubic γWW couplings. To construct those, it is good to start with the minimal electromagnetic couplings, obtained by the minimal substitution,

$$\partial_\mu \to D_\mu = \partial_\mu - ieA_\mu, \tag{8.6}$$

into equation (8.4b), leading to the following interaction Lagrangian (containing both cubic and quartic couplings):

[1] This description of the massive spin-1 particle is known as the *Proca model*.

$$\mathcal{L}_{\text{min}} = i\, e\, W^*_{\mu\nu} A^\mu W^\nu - i\, e\, A_\mu W^*_\nu W^{\mu\nu} + e^2 A^2\, W^*_\mu\, W^\mu. \tag{8.7}$$

The minimal substitution ensures the electromagnetic gauge symmetry, exhibited here by the invariance of the full Lagrangian under the following transformations:

$$A_\mu \to A_\mu + \partial_\mu \varphi, \qquad W^\mu \to e^{ie\varphi} W^\mu, \tag{8.8}$$

with φ an arbitrary scalar field. As the result, the charge and electromagnetic current are conserved.

In order to have arbitrary values of the magnetic and quadrupole moments, let us supplement the Lagrangian with the following terms:

$$\mathcal{L}_{\text{extra}} = i\, e\, \ell_1\, W^*_\mu W_\nu\, F^{\mu\nu} + \left(e\ell_2/2M^2\right)\left[(D^*_\mu W^*_\nu) W^\alpha \partial_\alpha F^{\mu\nu} + W^*_\alpha (D_\mu W_\nu)\partial^\alpha F^{\mu\nu}\right], \tag{8.9}$$

where ℓ_1 and ℓ_2 are the coupling constants, which contribute to the anomalous moments as follows:

$$\varkappa = \frac{1}{2}(\ell_1 - 1), \qquad \varsigma = \frac{1}{2}(1 + \ell_2 - \ell_1). \tag{8.10}$$

Thus, the natural values of the electromagnetic moments arise when $\ell_1 = 1$ and $\ell_2 = 0$. These are precisely the values corresponding with the Standard Model description of the charged gauge bosons, $W^\pm$. To understand why this choice of couplings is special, let us make a quick detour into the relation between gauge symmetries and the so-called degrees-of-freedom counting.

8.2 Gauge symmetries and spin degrees of freedom

Gauge symmetry is a divine principle of all the relativistic field theories of particles with spin-1 and higher, *viz*, 'higher-spin' field theories. The reason is that the number of field components exceeds the number of spin degrees of freedom (DoF), and the balance is established using gauge symmetries.

For example, a spin-1 particle ought to have three spin DoFs in the massive case and two in the massless[2]. A real four-vector field $W^\mu(x)$, describing it in a relativistic field theory, has four components. Writing the Lagrangian in a way analogous to the scalar theory, i.e. $\partial_\mu W_\nu \partial^\mu W^\nu$, would not work because all four components of the field act independently, giving rise to four spin DoFs; we need fewer. The correct Lagrangian is $\partial_\mu W_\nu \partial^\mu W^\nu - (\partial \cdot W)^2$, since it is invariant under a gauge transformation, $W^\mu \to W^\mu + \partial^\mu \varphi$, which ensures that the number of independent components is reduced to two. The mass term, $M^2 W^2$, raises this number to three, as it should.

The gauge symmetry thus eliminates the extra DoFs. The exact number of eliminated DoFs depends on the form of the gauge transformation. In general, the transformation can be written as $D_\mu \varphi^a(x)$, where D_μ is a differential operator of order d acting on the parameters of the gauge transformation φ^a, with $a = 1, \ldots, n$.

[2] In general, the number of spin DoFs is $2S + 1$ and 2 for, respectively, massive and massless particles with spin S.

The number of eliminated DoFs is then equal to[3] $(d + 1)n$. In the above example of a free spin-1 particle, the transformation $\partial_\mu\varphi$ has $d = 1 = n$, and hence the corresponding symmetry eliminates two DoFs.

Now, we can look at the theory considered in the previous section from the point of view of DoF counting. The free theory, equation (8.4), is alright: the kinetic terms of both A_μ and W_μ are gauge invariant. To make the bookkeeping easier, let us introduce A_μ^a, with $a = 1, 2, 3$, and define

$$A_\mu^1 = \frac{1}{\sqrt{2}}\left(W_\mu + W_\mu^*\right), \quad A_\mu^2 = \frac{i}{\sqrt{2}}\left(W_\mu - W_\mu^*\right), \quad A_\mu^3 = A_\mu. \tag{8.11}$$

The above free Lagrangian for the photon and $W^\pm$ is then written simply as

$$\mathcal{L} = -\frac{1}{4}\sum_{a=1}^{3} F_{\mu\nu}^a F^{a\mu\nu} - \frac{1}{2}M^2\sum_{a=1}^{2} A^a \cdot A^a, \qquad F_{\mu\nu}^a = \partial_\mu A_\nu^a - \partial_\nu A_\mu^a, \tag{8.12}$$

and the first term is invariant under the gauge transformation: $A_\mu^a \to A_\mu^a + \partial_\mu\varphi^a$.

The minimal substitution, equation (8.6), can now be written as $\partial_\mu A_\nu^a \to (\delta^{ab}\partial_\mu - e\varepsilon^{ab3}A_\mu^3)A_\nu^b$, where the summation over repeated indices is understood. Obviously, this substitution spoils some of the gauge symmetries of the free massless theory, retaining only the symmetry under

$$A_\mu^a \to A_\mu^a + (\delta^{a3}\partial_\mu + e\varepsilon^{ab3}A_\mu^b)\varphi^3. \tag{8.13}$$

The easiest way to restore the lost symmetries for $a = 1$ and 2 is to treat all the fields in the same fashion and do the minimal substitution as follows:

$$\partial_\mu A_\nu^a \to D_\mu^{ab}A_\nu^b, \qquad D_\mu^{ab} = \delta^{ab}\partial_\mu - e\varepsilon^{abc}A_\mu^c. \tag{8.14}$$

The resulting theory is none other than the SU(2) Yang–Mills theory, invariant under

$$A_\mu^a \to A_\mu^a + D_\mu^{ab}\varphi^b. \tag{8.15}$$

The more general minimal substitution, equation (8.14), leads to the following extra terms, in addition to the minimal coupling of equation (8.7):

$$\mathcal{L}_{\text{extraYM}} = i.\ e.\ W_\mu^* W_\nu F^{\mu\nu} + \frac{1}{2}e^2(|W \cdot W|^2 - |W|^4). \tag{8.16}$$

The first term is the cubic coupling, which fixes the magnetic and quadrupole moments of the W to the natural values (see equations (8.9) and (8.10)). The second term describes the W self-coupling, which emphasizes the non-Abelian nature of the Yang–Mills theory.

[3] A more rigorous formulation of this theorem (conveyed to me by Misha Vasiliev, who is world's greatest expert on higher-spin fields) can be found in [5].

To conclude, the consistent electrodynamics of a charged spin-1 particle is only realized within the Yang–Mills theory[4], where, due to the gauge symmetry, the electromagnetic moments are set to their natural values. The coupling (8.9) with arbitrary ℓ_1 and ℓ_2 is nevertheless useful in the sense of effective theory, and particularly in obtaining the low-energy theorems, which we now consider.

8.3 Tree-level unitarity: GDH sum rule

An important step towards deriving the sum rules is the forward Compton scattering. Recall from chapter 4 that the number of independent scalar amplitudes, in case of a real photon scattered off a massive spin-1 particle, is $2S + 2 = 4$. As the four Lorentz-covariant structures it is convenient to choose

$$\begin{aligned} \mathcal{O}_1 &= \varepsilon^* \cdot \varepsilon\, \chi^* \cdot \chi, & \mathcal{O}_2 &= \varepsilon^* \cdot \varepsilon\, q \cdot \chi^*\, q \cdot \chi, \\ \mathcal{O}_3 &= \varepsilon^* \cdot \chi^*\, \varepsilon \cdot \chi - \varepsilon^* \cdot \chi\, \varepsilon \cdot \chi^*, & \mathcal{O}_4 &= \varepsilon^* \cdot \chi^*\, \varepsilon \cdot \chi + \varepsilon^* \cdot \chi\, \varepsilon \cdot \chi^*, \end{aligned} \tag{8.17}$$

where ε and χ are the polarization vectors of, respectively, the photon and the spin-1 target; q is the photon four-momentum. The tree-level forward Compton-scattering amplitude computes (using equations (8.7) and (8.9)) as follows:

$$T^{\gamma W \to \gamma W}(\nu, t = 0) = 2M \sum_{i=1}^{4} \mathcal{A}_i(\nu) \mathcal{O}_i, \tag{8.18}$$

$$\begin{aligned} \mathcal{A}_1(\nu) &= -\frac{e^2}{M}, \\ \mathcal{A}_2(\nu) &= -\frac{e^2}{2M^3}\ell_2\,(1 + \ell_1) = -\frac{2e^2}{M^3}(1 + \varkappa)(\varkappa + \varsigma), \\ \mathcal{A}_3(\nu) &= -\frac{e^2\nu}{4M^2}(1 - \ell_1)^2 - \frac{e^2\nu^3}{16M^4}\ell_2^2 = -\frac{e^2\nu}{M^2}\varkappa^2 - \frac{e^2\nu^3}{4M^4}(\varkappa + \varsigma)^2, \\ \mathcal{A}_4(\nu) &= \frac{e^2\nu^2}{4M^3}\ell_2\,(1 - \ell_1) = -\frac{e^2\nu^2}{M^3}\varkappa(\varkappa + \varsigma), \end{aligned} \tag{8.19}$$

where $\nu = (s - M^2)/2M$ is the photon energy in the lab frame, $\mathcal{A}_i$ are scalar amplitudes.

The first amplitude represents the Thomson term. It is interesting that for the natural values ($\ell_1 = 1$, $\ell_2 = 0$), only the Thomson term survives, while the rest of the contributions vanish. In this case, the tree-level amplitude does not diverge at high energy and, therefore, is said to satisfy the 'tree-level unitarity'. The tree-level unitarity is a necessary condition for the theory to be renormalizable at one-loop level.

The various polarizabilities begin to contribute at order ν^2, and hence, as far as the sum rules involving only the electromagnetic moments are concerned, we need to focus on the terms of $O(1)$ and $O(\nu)$. The first one is the Thomson term (note that $\mathcal{O}_2$

[4] A better argumentation of this claim can be found in an insightful paper by Weinberg and Witten [6].

contains photon momenta and hence $\mathcal{A}_2$ contributes at $O(\nu^2)$), which, as already discussed, does not lead to a convergent sum rule. The term of $O(\nu)$, seen in the amplitude $\mathcal{A}_3$, is the anomalous magnetic moment contribution which leads to the GDH sum rule for the spin-1 case:

$$\boxed{\frac{2\alpha}{M^2}\varkappa^2 = -\frac{1}{2\pi^2}\int_0^\infty d\nu\, \frac{\sigma_0(\nu) - \sigma_2(\nu)}{\nu}.} \tag{8.20}$$

The quantity in the numerator on the right-hand side is the helicity-difference photoabsorption cross section. This form is basically the same as in the spin-1/2 case, see equation (4.18), which allows one to conjecture the following expression for the GDH sum rule for any spin S:

$$\boxed{\frac{2\alpha}{M^2}\varkappa^2 S = -\frac{1}{2\pi^2}\int_0^\infty d\nu\, \frac{\sigma_{1-S}(\nu) - \sigma_{1+S}(\nu)}{\nu},} \tag{8.21}$$

with the anomalous magnetic moment defined universally, for any spin, through the gyromagnetic ratio as $\varkappa = (g - 2)/2$.

The anomalous quadrupole moment starts to enter at $O(\nu^2)$ and cannot be disentangled from polarizability contributions; at least, not by means of sum rules derived from real Compton scattering. Conversely, some of the sum rules, which in the spin-1/2 case involved only polarizabilities, may now involve the electromagnetic moments. A remarkable example is provided by the sum rule for the forward spin polarizability (FSP) γ_0. Introducing it for a spin-1 particle in exactly the same way as for the spin-1/2 case results in the following sum rule [7]:

$$\boxed{-\frac{\alpha}{4M^4}(\varkappa + \varsigma)^2 + \gamma_0 = \frac{1}{4\pi^2}\int_0^\infty d\nu\, \frac{\sigma_0(\nu) - \sigma_2(\nu)}{\nu^3}.} \tag{8.22}$$

This has to be compared with the analogous spin-1/2 sum rule, equation (4.19), where the anomalous moment term is absent.

The FSP of the deuteron is conventionally defined to be a factor of two smaller [8] (i.e. γ_0 in equation (8.22) should be replaced with $2\gamma_0$). In addition, the contribution of anomalous moments is assumed to be absent from this sum rule. Fortunately, the contribution of the deuteron anomalous moments into this sum rule is numerically very small and may indeed be ignored[5]. The deuteron is, however, quite a peculiar system. It is a loose bound state of the proton and neutron, and as such, has rather large polarizabilities, enhanced by inverse powers of the binding momentum, which is two orders of magnitude smaller than the deuteron mass. For a 'more elementary'

[5] Using the currently known empirical values of the deuteron anomalous moments and mass (see [7] for details):

$$\varkappa_d \simeq -0.143, \qquad \varsigma_d \simeq 13.5, \qquad M_d \simeq 1.8756\ \text{GeV}, \qquad \alpha \simeq 1/137.036,$$

the correction due to the first term in equation (8.22) is about $-0.0263\ \text{GeV}^{-4} \simeq -4 \times 10^{-5}\ \text{fm}^4$, whereas γ_0 of the deuteron is expected to be of the order of a few fm^4, so, five orders of magnitude bigger. See [9] for a first experimental determination of the deuteron FSP.

particle, the interplay between the electromagnetic moments and the polarizability in the above sum rule can be entirely different.

8.4 Forward VVCS and virtual LbL scattering

As in the spin-1/2 case, considered in chapter 5, the doubly-virtual Compton scattering (VVCS) is the place to look for sum rules. The spin structure of the VVCS is of course more complicated for a spin-1 particle. The number of spin structures triples: six instead of two in the previous (spin-1/2) case[6]. The number of sum rules should increase accordingly. In the spin-1/2 case, the lowest order spin-dependent sum rules in the (quasi-)real-photon limit are GDH and Schwinger, so one per each independent spin structure function. Does the number of such sum rules for the spin-1 case triple as well?

A simple answer to this question is 'No.' There are not that many sum rules involving the electromagnetic moments without polarizabilities. A proper description of those sum rules requires us to dwell into the formalism *a la* chapter 5, except three times lengthier, which would certainly be too much for the 'concise' format we are bound to. In what follows we give only the highlights.

A good indication of sum rules for the electromagnetic moments of massive vector bosons can be obtained by looking at the sum rules for the $\gamma^*\gamma^*$ fusion, considered in the previous chapter. The forward LbL scattering is exactly analogous to the VVCS on massive vector bosons, provided the electromagnetic moments are set to zero, whilst the role of polarizabilities is played by the low-energy coefficients c_i from the effective Lagrangian (6.15), (6.16). The sum rules which do not involve polarizabilities appear in the LbL case as the *superconvergence* relations, i.e. the ones where the integral converges to zero.

Let us re-examine two of those relations:

$$\lim_{Q^2\to 0} \int_{\tilde{\nu}_0}^{\infty} d\tilde{\nu}\, \frac{1}{\tilde{\nu}} \tau_{\mathrm{TT}}^{a}(\tilde{\nu}, K^2, Q^2) = 0, \tag{8.23a}$$

$$\lim_{Q^2\to 0} \int_{\tilde{\nu}_0}^{\infty} d\tilde{\nu}\, \frac{1}{Q} \tau_{\mathrm{TL}}^{a}(\tilde{\nu}, K^2, Q^2) = 0, \tag{8.23b}$$

where τ_{TT}^{a} and τ_{TL}^{a} are cross section quantities describing the fusion of two virtual photons into everything ($\gamma^*\gamma^*\to X$), see [10].[7] The photons have the space-like virtualities K^2 and Q^2, respectively, the invariant $\tilde{\nu} = (s + K^2 + Q^2)/2$, and $\tilde{\nu}_0$ is the inelastic threshold. By definition, $\tau_{\mathrm{TT}}^{a} = (\sigma_0 - \sigma_2)/2$, and hence the first relation is identified as the GDH sum rule in the absence of electromagnetic moments. To see

[6] Recall from chapter 4 that the number of independent forward VVCS amplitudes is $5S + 3$ for bosons (with nonvanishing spin), and $5S + 3/2$ for fermions. For any S, two of these amplitudes are spin independent. Hence, the number of spin structure functions is 6 for $S = 1$, and 2 for $S = 1/2$.

[7] The response function τ_{TL}^{a} is interchanged with τ_{TL} in [11], as compared with [10], because of a different convention for the longitudinal polarization vector: $\varepsilon_0^* = \varepsilon_0$ versus $\varepsilon_0^* = -\varepsilon_0$, respectively. Here we assume the latter convention.

that the other one is the analogue of the Schwinger sum rule, let us introduce the spin-1 structure functions, if only very schematically.

The structure function formalism, analogous to the spin-1/2 case, was developed in the context of the deuteron by Hoodbhoy, Jaffe, and Manohar [12]. The Lorentz-covariant form of the forward VVCS amplitude splits again into the symmetric and antisymmetric parts, $T^{\mu\nu} = T_S^{\mu\nu} + T_A^{\mu\nu}$, which contain six and two amplitudes, respectively. The imaginary parts of those amplitudes are given by the following eight structure functions,

$$\text{scalar: } F_1(x, Q^2), \;\; F_2(x, Q^2), \tag{8.24a}$$

$$\text{vector: } g_1(x, Q^2), \;\; g_2(x, Q^2), \tag{8.24b}$$

$$\text{tensor: } b_i(x, Q^2), \;\; i = 1, \dots, 4. \tag{8.24c}$$

They are referred to, respectively, as the scalar, vector, and tensor structure functions. The latter are new compared to the spin-1/2 case. They involve the longitudinal polarization of the spin-1 target. Just as the scalar ones, the tensor structure functions enter through the symmetric part of the VVCS tensor, whilst the vector ones through the antisymmetric part.

These structure functions can be matched to the cross section quantities,

$$\sigma_{\mathrm{TT}}, \; \sigma_{\mathrm{TL}}, \; \sigma_{\mathrm{LT}}, \; \sigma_{\mathrm{LL}}, \; \tau_{\mathrm{TT}}, \; \tau_{\mathrm{TL}}, \; \tau_{\mathrm{TT}}^a, \; \tau_{\mathrm{TL}}^a, \tag{8.25}$$

used to describe the $\gamma^*\gamma^*$ collisions in chapter 7. Of course, one needs to keep in mind that one of the virtual photons in the LbL case corresponds with the massive boson, hence, the virtuality must match the mass: $K^2 = -M^2$. The energy invariants are matched as $\tilde{\nu} = M\nu = Q^2/2x$. The precise matching of the structure functions can be worked out by comparing the helicity amplitudes for the two cases. As the result we, in particular, have

$$g_1 \sim \frac{Q}{\nu}\tau_{\mathrm{TL}}^a + \tau_{\mathrm{TT}}^a, \tag{8.26a}$$

$$g_2 \sim \frac{\nu}{Q}\tau_{\mathrm{TL}}^a - \tau_{\mathrm{TT}}^a, \tag{8.26b}$$

and the analogy with the spin-1/2 case becomes apparent, see equation (5.4). The superconvergence relation (8.23b) thus corresponds with the Schwinger sum rule.

To show this more explicitly, let us cast the above superconvergence relations into sum rules for spin structure functions of a real photon, denoted as $g_i^\gamma(x, Q^2)$. For this, we interchange the virtualities, $K^2 \leftrightarrow Q^2$, such that K^2 is set to zero whereas Q^2 is arbitrary. The exact relation between g_i^γ and τ in this case becomes:

$$g_1^\gamma = \frac{1}{4\pi^2\alpha} \lim_{K^2 \to 0} \frac{\tilde{\nu}}{\sqrt{1 - K^2Q^2/\tilde{\nu}}}\left(\frac{KQ}{\tilde{\nu}}\tau_{\mathrm{TL}}^a + \tau_{\mathrm{TT}}^a\right), \tag{8.27a}$$

$$g_2^\gamma = \frac{1}{4\pi^2\alpha} \lim_{K^2\to 0} \frac{\tilde{\nu}}{\sqrt{1 - K^2Q^2/\tilde{\nu}^2}} \left(\frac{\tilde{\nu}}{KQ} \tau_{\mathrm{TL}}^a - \tau_{\mathrm{TT}}^a \right). \tag{8.27b}$$

Solving these for τ and substituting in equation (6.20) with $\tilde{\nu} = Q^2/2x$ we, respectively, have

$$\boxed{\int_0^{x_0} dx\, g_1^\gamma(x, Q^2) = 0} \tag{8.28a}$$

$$\boxed{\int_0^{x_0} dx \left[g_1^\gamma(x, Q^2) + g_2^\gamma(x, Q^2) \right] = 0}. \tag{8.28b}$$

Given that the elastic contribution is absent in the case of LbL scattering, we can extend the upper limit of integration to unity, and thus these two sum rules imply as well the BC sum rule: $\int_0^1 dx\, g_2^\gamma(x, Q^2) = 0$.

The third superconvergence relation (7.23b), derived originally in [11], has no spin-1/2 analogue. When extended to the massive boson case, it may, potentially, yield a sum rule for the quadrupole moment. Some work in this direction has been done, e.g. for the tensor structure function b_1, see [13] and references therein. Yet, a sum rule involving a quadrupole moment (and no polarizabilities!) has not been written down. For now, this task is given away as an exercise to this chapter.

References

[1] Weinberg S 1970 *Lectures on Elementary Particles and Quantum Field Theory* (Brandeis University Summer Institute 1970 vol 1) (Cambridge, MA: MIT Press)

[2] Ferrara S, Porrati M and Telegdi V L 1992 $g = 2$ as the natural value of the tree level gyromagnetic ratio of elementary particles *Phys. Rev.* D **46** 3529–37

[3] Alexandrou C, Korzec T, Koutsou G, Lorce C, Negele J W, Pascalutsa V, Tsapalis A and Vanderhaeghen M 2009 Quark transverse charge densities in the Δ(1232) from lattice QCD *Nucl. Phys.* A **825** 115–44

[4] Lorce C 2009 Electromagnetic properties for arbitrary spin particles: natural electromagnetic moments from light-cone arguments *Phys. Rev.* D **79** 113011

[5] Pascalutsa V and Timmermans R 1999 Field theory of nucleon to higher-spin baryon transitions *Phys. Rev.* C **60** 042201

[6] Weinberg S and Witten E 1980 Limits on massless particles *Phys. Lett.* B **96** 59–62

[7] Hagelstein F 2014 Sum rules for electromagnetic moments and polarizabilities of spin-1 particles in massive Yang–Mills QED *Master's thesis* University of Mainz

[8] Ji X and Li Y 2004 Sum rules and spin dependent polarizabilities of the deuteron in effective field theory *Phys. Lett.* B **591** 76–82

[9] Ahmed M W, Blackston M A, Perdue B A, Tornow W, Weller H R, Norum B, Sawatzky B, Prior R M and Spraker M C 2008 Near-threshold deuteron photodisintegration: an indirect determination of the Gerasimov–Drell–Hearn sum rule and forward spin polarizability (γ_0) for the deuteron at low energies *Phys. Rev.* C **77** 044005

[10] Budnev V, Ginzburg I, Meledin G and Serbo V 1975 The two photon particle production mechanism. Physical problems. Applications. Equivalent photon approximation *Phys. Rep.* **15** 181–281

[11] Pascalutsa V, Pauk V and Vanderhaeghen M 2012 Light-by-light scattering sum rules constraining meson transition form factors *Phys. Rev.* D **85** 116001

[12] Hoodbhoy P, Jaffe R L and Manohar A 1989 Novel effects in deep inelastic scattering from spin 1 hadrons *Nucl. Phys.* B **312** 571–88

[13] Close F E and Kumano S 1990 A sum rule for the spin dependent structure function $b - 1(x)$ for spin one hadrons *Phys. Rev.* D **42** 2377–9

IOP Publishing

Causality Rules (Second Edition)
Dispersion theory in non-elementary particle physics
Vladimir Pascalutsa

Chapter 9

Vacuum polarization and $g - 2$ of the muon

Vacuum polarization is one of the most fundamental quantum phenomena. It describes the modification of the photon propagation by quantum fluctuations of the vacuum. Vacuum polarization affects many interesting quantities in physics: atomic spectra (the Lamb shift and hyperfine structure), the anomalous magnetic moment of the electron and muon, and the running of the fine-structure constant α with energy. These quantities are measured very precisely, providing stringent tests of the quantum theory and of the Standard Model (SM) in particular. Some of these tests will be explained briefly here, after an extensive introduction to the general properties of vacuum polarization leading to a dispersion relation and sum rules.

9.1 Vacuum polarization in QED

The Heisenberg uncertainty principle of quantum theory allows for brief fluctuations of energy and therefore the appearance and disappearance of virtual particles without violating the conservation of energy. A photon propagating through the vacuum is affected by these quantum fluctuations, most notably by the production and annihilation of electrically charged particle.

The Feynman diagram in figure 9.1 represents such an effect due to a virtual particle–antiparticle pair, e.g. the electron–positron (e^+e^-) pair. A calculation of this diagram is one of the most basic exercises in quantum electrodynamics (QED). Let us sketch this calculation, without going into details (the missing steps can be found in textbooks):

$$\begin{aligned}\Pi^{\mu\nu}(q) &= i\int \frac{d^4k}{(2\pi)^4}\mathrm{tr}\left[e\gamma^\mu \frac{\not{k}+m}{k^2-m^2+i0^+}e\gamma^\nu \frac{\not{q}+\not{k}+m}{(q+k)^2-m^2+i0^+}\right]\\ &= \frac{e^2}{2\pi^2}(q^2g^{\mu\nu}-q^\mu q^\nu)\left\{L_{\mathrm{UV}}+\int_0^1 dx\, x(1-x)\log\left[m^2-x(1-x)q^2-i0^+\right]\right\},\end{aligned} \tag{9.1}$$

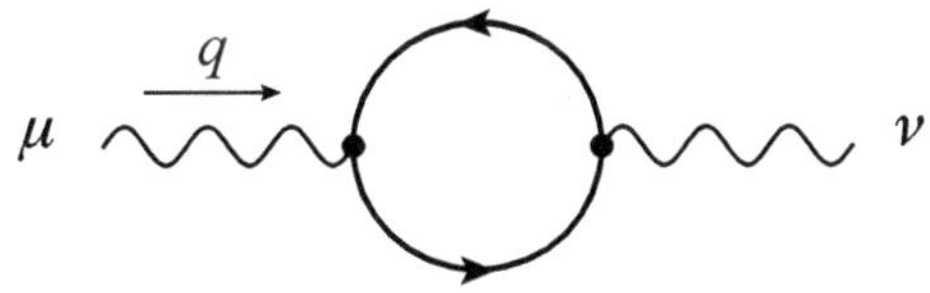

Figure 9.1. One-loop vacuum polarization, the leading quantum correction to the photon propagator.

where e and m are the lepton change and mass, and the final expression contains an infinite constant L_{UV} which represents the ultra-violet divergence of this diagram. We first observe that the vacuum polarization, in momentum space, has the following general form

$$\Pi^{\mu\nu}(q) = (g^{\mu\nu}q^2 - q^\mu q^\nu)\Pi(q^2), \tag{9.2}$$

where q^μ is the photon four-momentum. In addition to Lorentz covariance, this form is dictated by the electromagnetic gauge invariance, which expresses here the transversality condition: $q_\mu \Pi^{\mu\nu}(q) = q_\nu \Pi^{\mu\nu}(q) = 0$. Hence, it is entirely characterized by a single scalar function of q^2, which in our example is given by

$$\Pi(q^2) = \frac{2\alpha}{\pi}\left\{L_{\mathrm{UV}} + \int_0^1 dx\, x(1-x)\log\left[1 - x(1-x)\frac{q^2}{m^2} - i0^+\right]\right\}. \tag{9.3}$$

The imaginary part of this function stems from the logarithm and is found as follows[1]:

$$\begin{aligned}\operatorname{Im}\Pi(q^2) &= \frac{2\alpha}{\pi}\int_0^1 dx\, x(1-x)\operatorname{Im}\log\left[1 - x(1-x)\frac{q^2}{m^2} - i0^+\right]\\ &= \{x \to x + \tfrac{1}{2}\} = \frac{2\alpha}{\pi}\int_{-\frac{1}{2}}^{\frac{1}{2}} dx\left(\frac{1}{4} - x^2\right)\operatorname{Im}\log\left[1 - \left(\frac{1}{4} - x^2\right)\frac{q^2}{m^2} - i0^+\right]\\ &= \{v \equiv \sqrt{1 - 4m^2/q^2}\} = \frac{2\alpha}{\pi}\int_{-\frac{1}{2}v}^{\frac{1}{2}v} dx\left(\frac{1}{4} - x^2\right)(-\pi)\\ &= -\frac{1}{2}\alpha v\left(1 - \frac{1}{3}v^2\right).\end{aligned} \tag{9.4}$$

We are now in a position to verify the dispersion relation for the vacuum polarization.

[1] In the case of scalar QED (e.g. $\pi^+\pi^-$ pair), we would obtain

$$\operatorname{Im}\Pi(v) = -\frac{\alpha}{12}v^3, \quad \text{with } v = \sqrt{1 - 4m^2/q^2}.$$

Considering $\Pi(s)$, with $s = q^2$, in the complex plane of s we find that its analytic properties must be exactly the same as for the function $f(\omega)$, considered earlier in chapter 2. The function analytic everywhere apart for the branch cut on the positive real axis (see figure 2.1), which represents the possibility for a real (rather than virtual) particle production when the photon has time-like virtuality ($q^2 > 0$). We thus arrive at the unsubtracted dispersion relation, obtained under the usual convergence assumption,

$$\Pi(s) = \frac{1}{\pi}\int_{s_0}^{\infty} ds' \, \frac{\operatorname{Im}\Pi(s')}{s' - s - i0^+}, \tag{9.5}$$

where s_0 is the particle-production threshold. However, substituting the imaginary part that we have just found in one-loop QED, i.e.

$$\operatorname{Im}\Pi(s') = -\frac{\alpha}{3}\left(1 + \frac{2m^2}{s'}\right)\sqrt{1 - \frac{4m^2}{s'}}, \tag{9.6}$$

we see that the integral diverges at large s'. On one hand, this is consistent with having the ultra-violet divergence L_{PUV} in the original expression. On the other hand, it apparently invalidates the convergence condition that leads to the unsubtracted dispersion relation.

Making a subtraction at $s = 0$, we obtain the once-subtracted dispersion relation,

$$\Pi(s) = \Pi(0) + \frac{s}{\pi}\int_{s_0}^{\infty} \frac{ds'}{s'} \, \frac{\operatorname{Im}\Pi(s')}{s' - s - i0^+}, \tag{9.7}$$

but then we need to know something about $\Pi(0)$. For this we consider the impact it has on the photon propagator. In the absence of vacuum polarization, the photon propagator has the form (in the Feynman gauge)

$$\Delta_{\mu\nu}(q) = \frac{-ig_{\mu\nu}}{q^2 + i0^+}, \tag{9.8}$$

while a resummation of the vacuum-polarization contributions, represented by figure 9.2, results in the following modification:

$$\tilde{\Delta}_{\mu\nu}(q) = \Delta_{\mu\nu}(q) + \Delta_{\mu\rho}(q)\, i\Pi^{\rho\sigma}\, \Delta_{\sigma\nu}(q) + \cdots = \frac{\Delta_{\mu\nu}(q)}{1 - \Pi(q^2)}. \tag{9.9}$$

The important point here is that at the real photon point, $q^2 = 0$, the propagator should not be modified, i.e. $\tilde{\Delta}_{\mu\nu}(q) \to \Delta_{\mu\nu}(q)$, and hence

$$\Pi(q^2 \to 0) = 0. \tag{9.10}$$

Figure 9.2. The photon propagator 'dressed' by the vacuum-polarization contributions. The blob with 1PI stands for the one-particle-irreducible graphs.

This is the on-shell renormalization condition of the electromagnetic field[2]. It ensures that we observe a massless photon with probability 1.

Under this renormalization condition, the vacuum polarization satisfies the following dispersion relation

$$\boxed{\Pi(s) = \frac{s}{\pi}\int_{s_0}^{\infty}\frac{ds'}{s'}\,\frac{\operatorname{Im}\Pi(s')}{s' - s - i0^+}}. \tag{9.11}$$

It is easily verifiable in our QED example by substituting $s_0 = 4m^2$ and $\operatorname{Im}\Pi$ of equation (9.6) to the right-hand side, and recovering the expression of $\Pi(s) - \Pi(0)$ from (9.3).

9.2 Unitarity and sum rules

The next step is to use unitarity that relates the discontinuity across the branch cut to the probability of particle production. This probability is given essentially by the total decay width, $\Gamma_{\gamma^*\to X}$, which describes the transition of the virtual (time-like) photon, γ^*, into any bunch of real particles, denoted as X. The analogue of the optical theorem for the scattering amplitudes is, in this case, given by

$$\operatorname{Im}\Pi(s) = -\frac{1}{\sqrt{s}}\,\Gamma_{\gamma^*\to X}(s). \tag{9.12}$$

This width cannot be measured directly, since a virtual photon is only an intermediate particle. In a laboratory the time-like photon is usually created at e^+e^- colliders. What is measured then is the cross section of $e^+e^-\to X$. This process is dominated by the exchange of one photon, i.e. $e^+e^-\to\gamma^*\to X$, which is very close to what we need. In this one-photon-exchange approximation, the cross section factorizes into the probability of the photon transition into the e^+e^- pair and the sought probability of $\gamma^*\to X$:

$$\sigma_{e^+e^-\to\gamma^*\to X}(s) = \frac{1}{s^2}\Gamma_{\gamma^*\to e^+e^-}(s)\,\Gamma_{\gamma^*\to X}(s) = \frac{1}{s}\operatorname{Im}\Pi_{e^+e^-}(s)\operatorname{Im}\Pi_X(s). \tag{9.13}$$

In this way, we can assess the contribution of a given channel X through its production in e^+e^- collisions, given that $\Pi_{e^+e^-}$ can be calculated in QED.

It is especially useful for assessing the *hadronic contributions* to vacuum polarization, referred to as hadronic vacuum polarization (HVP), by using the experimentally measured cross sections of $e^+e^-\to$ hadrons. Since the production of hadrons is measured at energies much higher than the electron mass, $s \gg m_e^2$, the

[2] By gauge invariance, this condition is related to the charge renormalization. As the photon is mediating the electromagnetic interaction between two charges, the effect of vacuum polarization leads effectively to a modification of the QED coupling:

$$\alpha^{\text{eff.}}(q^2) = \frac{\alpha}{1 - \Pi(q^2)}.$$

The MUonE Project experiment, currently proposed at CERN, plans to measure precisely this modification in high-energy muon-electron scattering.

mass can be neglected in $\mathrm{Im}\,\Pi_{e^+e^-}$, which is given by equation (9.6). Expressing the HVP from (9.13) and the dispersion relation (9.7) we thus obtain

$$\mathrm{Im}\,\Pi_{\mathrm{had}}(s) = -\frac{\alpha}{3s}\,\sigma_{e^+e^-\to\,\mathrm{had}}(s), \tag{9.14a}$$

$$\Pi_{\mathrm{had}}(s) = -\frac{\alpha s}{3\pi}\int_{s_0}^{\infty}\frac{ds'}{s'^{\,2}}\,\frac{\sigma_{e^+e^-\to\,\mathrm{had}}(s')}{s' - s - i0^+}, \tag{9.14b}$$

where s_0 is set at the pion-production threshold. Remember that these relations hold at leading order in α only. To obtain the higher-order corrections we would need to compute $\Pi_{e^+e^-}$ to two loops and, more importantly, consider the two-photon-exchange e^+e^- collisions.

Historically, this procedure had been implemented somewhat differently. One introduces the following ratio of cross sections,

$$R_\gamma^{\mathrm{had}}(s) = \frac{\sigma_{e^+e^-\to\gamma^*\to\,\mathrm{had}}(s)}{\sigma_{e^+e^-\to\gamma^*\to\mu^+\mu^-}(s)}. \tag{9.15}$$

Then, using equation (9.13), the HVP is expressed as

$$\mathrm{Im}\,\Pi_{\mathrm{had}}(s) = \mathrm{Im}\,\Pi_{\mu^+\mu^-}(s)\;R_\gamma^{\mathrm{had}}(s), \tag{9.16a}$$

$$\Pi_{\mathrm{had}}(s) \approx -\frac{\alpha s}{3\pi}\int_{s_0}^{\infty}\frac{ds'}{s'}\,\frac{R_\gamma^{\mathrm{had}}(s')}{s' - s - i0^+}, \tag{9.16b}$$

where in the last expression we have again neglected the mass in the lepton vacuum polarization. Note, however, that now it is the muon mass, which is not necessarily much smaller than the lightest hadronic scales. Recall that the muon and the pion masses are comparable.

Figure 9.3 represents a summary of the empirical information on the ratio (9.15). It required an impressive experimental and theoretical effort to establish this quantity over such a wide energy region. It emphasizes how important this observable is, and how important is the accurate determination of the HVP contributions. Below we consider one of the most interesting examples, at least at the present time.

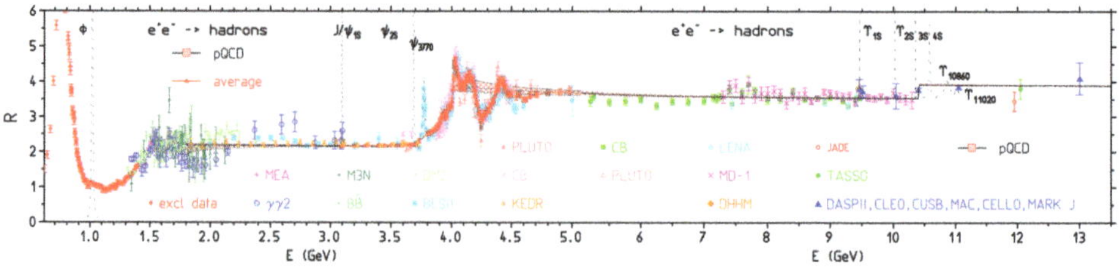

Figure 9.3. The ratio of electron–positron annihilation into hadrons and electron–positron annihilation into a muon pair. (Reproduced with permission from [3]. Copyright 2020 Elsevier.)

9.3 Introduction to the muon anomaly

As we already noted a number of times, the use of dispersion relations to assess the hadronic effects is especially appealing in the low-energy domain, where *ab initio* QCD calculations are very complicated and lagging behind.

A prominent and timely example is provided by the anomalous magnetic moment of the muon,

$$a_\mu = \frac{1}{2}(g_\mu - 2), \tag{9.17}$$

where g_μ is the muon gyromagnetic ratio, which is the proportionality factor between the magnetic moment $\vec{M}$ and the spin $\vec{S}$:

$$\vec{M}_\mu = g_\mu \frac{e}{2m_\mu} \vec{S}. \tag{9.18}$$

At the classical level, $g_\mu = 2$. However, quantum corrections give rise to non-vanishing a_μ. Recall that the leading one-loop correction is given by (see figure 5.6 in chapter 5)

$$a_\mu^{\text{one-loop}} = \frac{\alpha}{2\pi} \simeq 1\,161\,410 \times 10^{-9}, \tag{9.19}$$

which was first obtained by Schwinger in 1949 [1]. Since then, the knowledge of this quantity has been tremendously improved both theoretically and experimentally (see [2, 3] for comprehensive reviews). Currently there is an interesting discrepancy between theory and experiment, as seen in figure 9.4. The experimental result has just been improved once more [4], reducing the experimental uncertainty to 0.2 parts-per-million (ppm), and reaffirming the discrepancy with the Standard Model.

The largest systematic uncertainty of the Standard Model calculation stems from the hadronic contributions. Here one distinguishes two topologies: HVP and hadronic light-by-light scattering (HLbL), see figure 9.5. The HVP is the leading hadronic contribution. The HLbL contains one extra photon and an extra loop, hence is suppressed by at least a factor of $\alpha \sim 10^{-2}$, as compared to HVP. On the other hand, the two effects contribute nearly equally to the theoretical uncertainty, because the HLbL contribution is poorly known (worse than 20% relative precision), whereas the claimed accuracy of dispersive data-driven evaluations of the HVP contribution to a_μ has reached the subpercent level [3, 6].

It is important to point out that, at the time of writing this text, many of the lattice-QCD calculations of HVP contribution (e.g. [7]), by different collaborations, disagree with the state-of-art data-driven evaluations. Using only these latest lattice-QCD results for HVP would basically resolve the discrepancy, seen before in figure 9.4, between the Standard Model and experiment. It will be very interesting to see how this field develops further. Let us, in the meantime, consider the method of the data-driven evaluations in somewhat more detail.

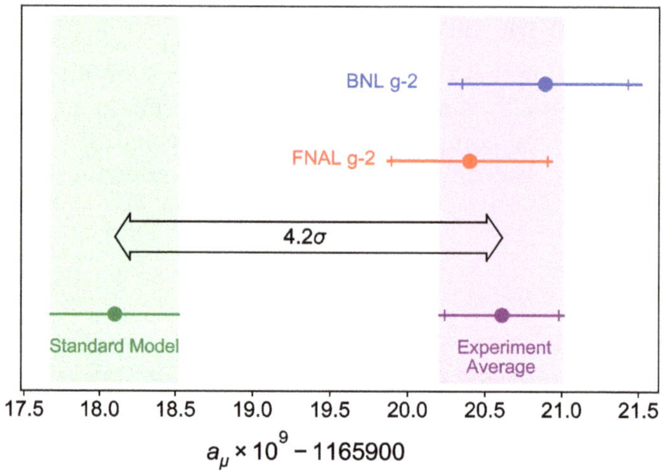

Figure 9.4. From top to bottom: experimental values of a_μ from the Brookhaven (BNL) E821 experiment, and the recent Fermilab (FNAL) experiment, together with their combined average. The inner tick marks indicate the statistical contribution to the total uncertainties. The Muon $g-2$ Theory Initiative recommended value [3] for the Standard Model is also shown. (Reproduced with permission from [5]. Copyright 2021 the American Physical Society.)

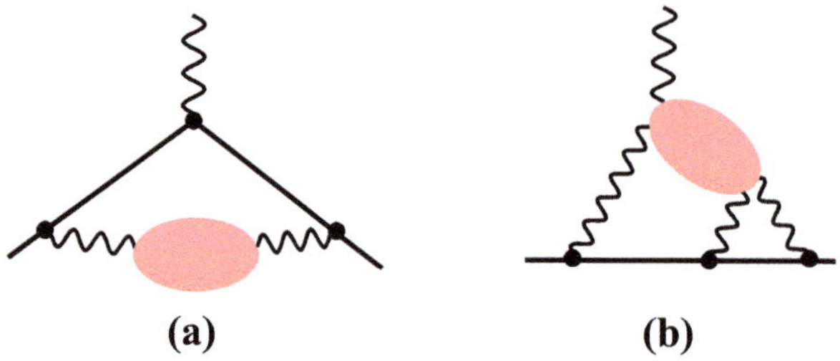

Figure 9.5. Hadronic corrections to the anomalous magnetic moment: (a) hadronic vacuum polarization and (b) hadronic light-by-light scattering.

9.4 Hadronic vacuum polarization in the muon anomaly

Consider the vacuum-polarization contribution in figure 9.5(a). The photon line between the muon lines corresponds to the second term in the expansion of the 'dressed' photon propagator in figure 9.2

$$\Delta_{\mu\rho}(q)\, i\Pi^{\rho\sigma}\, \Delta_{\sigma\nu}(q) = \frac{-ig^{\mu\nu}}{q^2 + i0^+}\Pi(q^2) = -\frac{1}{\pi}\int_{s_0}^{\infty}\frac{ds}{s}\frac{-ig^{\mu\nu}}{q^2 - s + i0^+}\,\mathrm{Im}\,\Pi(s), \quad (9.20)$$

where in the last step we have used the dispersive representation of the vacuum polarization (9.7). We can see that, as far as the loop integration over the four-momentum q is concerned, the photon propagator becomes massive. The massless photon propagator $1/q^2$ is, in the presence of vacuum polarization, being replaced with $1/(q^2 - s)$. The loops diagram can be easily performed for a massive photon, with variable mass $\sqrt{s}$. The integral over s can be performed afterwards. This is a typical idea behind evaluating the vacuum-polarization contribution to various quantities.

The one-loop contribution to a_μ, in the case of the photon with mass $\sqrt{s}$, is given by

$$\frac{\alpha}{\pi}K(s/m_\mu^2) = \frac{\alpha}{\pi}\int_0^1 dx\, \frac{x^2(1-x)}{x^2 + (1-x)s/m_\mu^2}. \tag{9.21}$$

As a check, for the massless photon ($s \to 0$), we recover $\alpha/2\pi$. The entire contribution of figure 9.5(a) can thus be written as a sum rule:

$$a_\mu = -\frac{\alpha}{\pi^2}\int_{s_0}^\infty \frac{ds}{s}\, K(s/m_\mu^2)\, \mathrm{Im}\, \Pi(s), \tag{9.22}$$

with $\mathrm{Im}\,\Pi(s)$, in the case of the HVP contribution, given by equation (9.16a) in terms of the cross section ratio R_γ^{had} defined in (9.15):

$$a_\mu(\mathrm{HVP}) = -\frac{\alpha}{\pi^2}\int_{s_0}^\infty \frac{ds}{s}\, K(s/m_\mu^2)\, \mathrm{Im}\,\Pi_{\mu^+\mu^-}(s)\; R_\gamma^{\mathrm{had}}(s) \tag{9.23}$$

$$\approx \frac{\alpha^2}{3\pi^2}\int_{s_0}^\infty \frac{ds}{s}\, K(s/m_\mu^2)\, R_\gamma^{\mathrm{had}}(s), \tag{9.24}$$

where in the last step we have made the usual approximation, $m_\mu^2 \ll s$ in equation (9.6), which results in $\mathrm{Im}\,\Pi_{\mu^+\mu^-} = -\frac{1}{3}\alpha$.

9.5 Muon anomaly via the Schwinger sum rule

In the following, we will show that the same data-driven formula for the leading HVP contribution can be derived from the Schwinger sum rule introduced in chapter 5, see (5.22). The empirical input entering the Schwinger sum rule are the muon spin structure functions, or equivalently, the longitudinal-transverse photoabsorption cross section of the muon. The mechanism of hadron photo-production off the muon through time-like Compton scattering, shown in figure 9.6(a), is the relevant one for the HVP topology.

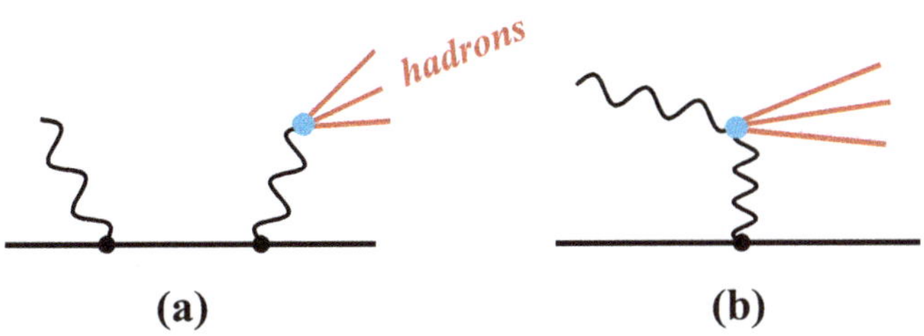

Figure 9.6. Mechanisms of hadron photo-production off a lepton: (a) time-like Compton scattering (crossed diagram omitted) and (b) Primakoff effect.

Let us factorize the invariant mass distribution $d\sigma(\gamma\mu \to \mu X)/dM_X^2$, arising from figure 9.6(a), into the cross sections of time-like Compton scattering $\sigma(\gamma\mu \to \gamma^*\mu)$ and of the subsequent photon decay into hadrons $\Gamma(\gamma^* \to X)$. The cross section of time-like Compton scattering at tree level, see figure 5.5, with initial and final photon virtualities $q^2 = -Q^2$ and $q'^2 = M_X^2$, follows from spinor QED as

$$\left[\frac{\sigma_{\rm LT}^{\gamma\mu\to\gamma^*\mu}(\nu, Q^2)}{Q}\right]_{Q^2=0} = \frac{\pi\alpha^2}{2m^2\nu^3}\left[-(5s + m^2 + M_X^2)\lambda + (s + 2m^2 - 2M_X^2)\log\frac{\beta+\lambda}{\beta-\lambda}\right], \quad (9.25)$$

with the following kinematic quantities,

$$\begin{aligned} s &= m_\mu^2 + 2m_\mu\nu, \\ \beta &= \frac{s + m_\mu^2 - M_X^2}{2s}, \\ \lambda &= \frac{1}{2s}\sqrt{\left[s - (m_\mu + M_X)^2\right]\left[s - (m_\mu - M_X)^2\right]}. \end{aligned} \quad (9.26)$$

In the limit of $M_X \to 0$, one obtains again the cross section for virtual Compton scattering in equation (5.24). The virtual photon decay is, by means of unitarity, expressed via $\mathrm{Im}\,\Pi^{\rm had}(q'^2)$. Plugging a factorized cross section for the mechanism in figure 9.6(a) into the Schwinger sum rule,

$$\begin{aligned} a_\mu &= \frac{m_\mu^2}{\pi^2\alpha}\int_{s_0}^\infty dM_X^2\int_{\nu_0}^\infty d\nu\left[\frac{1}{Q}\frac{d\sigma_{\rm LT}^{\gamma\mu\to\mu X}(\nu, Q^2)}{dM_X^2}\right]_{Q^2=0} \\ &= -\frac{1}{\pi}\int_{s_0}^\infty dM_X^2\,\frac{\mathrm{Im}\,\Pi^{\rm had}(M_X^2)}{M_X^2}\frac{m_\mu^2}{\pi^2\alpha}\int_{\nu_0}^\infty d\nu\left[\frac{\sigma_{\rm LT}^{\gamma\mu\to\gamma^*\mu}(\nu, Q^2)}{Q}\right]_{Q^2=0}, \end{aligned} \quad (9.27)$$

where $\nu_0 = M_X(1 + M_X^2/2m_\mu^2)$ is the photo-production threshold, and performing the integration over ν, we obtain the kernel function of equation (9.21):

$$\frac{m_\mu^2}{\pi^2\alpha}\int_{\nu_0}^\infty d\nu\left[\frac{\sigma_{\rm LT}^{\gamma\mu\to\gamma^*\mu}(\nu, Q^2)}{Q}\right]_{Q^2=0} = \frac{\alpha}{\pi}K(M_X^2/m_\mu^2). \quad (9.28)$$

Thus, replacing M_X^2 by s, we reproduce the sum rule for the HVP contribution to the muon anomalous magnetic moment (see (9.22)):

$$a_\mu = -\frac{\alpha}{\pi^2}\int_{s_0}^\infty \frac{ds}{s}\,K(s/m_\mu^2)\,\mathrm{Im}\,\Pi(s). \quad (9.29)$$

Now, what about the Primakoff mechanism, shown in figure 9.6(b)? It adds up incoherently with figure 9.6(a) (i.e. there is no interference term), because of C-parity conservation, *viz*, Furry's theorem.

Both mechanisms begin to contribute at $O(\alpha^3)$ to $\sigma_{\rm LT}$ and hence at $O(\alpha^2)$ to $g - 2$. As we have just seen, the time-like Compton corresponds with the HVP contribution. Hence, the Primakoff mechanism to $g - 2$ should vanish exactly, as it does not

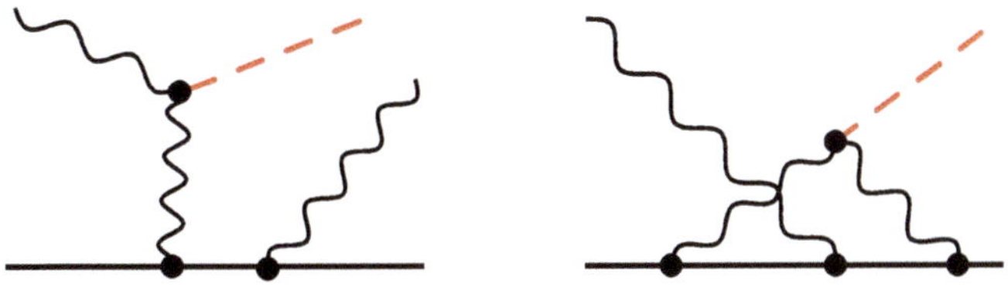

Figure 9.7. Subleading mechanisms accompanying the single-meson photo-production.

correspond with the HVP contribution, and there is no other hadronic contribution to $g - 2$ at this order. An explicit proof of this statement involves the light-by-light scattering sum rules, considered in chapter 7. As the result, the Primakoff mechanism can only contribute in interference with subleading effects, such as the one given by the second diagram of figure 9.7 for the case of π^0 production. In this case it corresponds with the HLbL contribution.

References

[1] Schwinger J 1949 Quantum electrodynamics. II. Vacuum polarization and self-energy *Phys. Rev.* **75** 651–79

[2] Jegerlehner F 2017 *The Anomalous Magnetic Moment of the Muon* (Springer Tracts in Modern Physics vol 274) (Berlin: Springer)

[3] Aoyama T *et al* 2020 The anomalous magnetic moment of the muon in the standard model *Phys. Rep.* **887** 1–166

[4] Aguillard D P *et al* 2023 Measurement of the positive muon anomalous magnetic moment to 0.20 ppm arXiv: 2308.06230

[5] Abi B *et al* 2021 Measurement of the positive muon anomalous magnetic moment to 0.46 ppm *Phys. Rev. Lett.* **126** 141801

[6] Davier M 2017 Update of the hadronic vacuum polarisation contribution to the muon $g - 2$ *Nucl. Part. Phys. Proc.* **287–288** 70–5

[7] Borsanyi S *et al* 2021 Leading hadronic contribution to the muon magnetic moment from lattice QCD *Nature* **593** 51–5

IOP Publishing

Causality Rules (Second Edition)
Dispersion theory in non-elementary particle physics
Vladimir Pascalutsa

Chapter 10

Dispersion theory of hydrogen-like atoms

This chapter demonstrates the use of the dispersion technique in obtaining various corrections to the Lamb shift and hyperfine structure of hydrogen-like atoms. Many of the corrections considered here, such as the electron vacuum-polarization, or the nuclear polarizability effects, are especially prominent in muonic hydrogen and other light muonic atoms. Muonic hydrogen is the bound state of the negative muon (μ^-) with the proton (p), rather than the electron–proton of normal hydrogen. Because the muon mass is about a factor of 200 larger than the electron mass, the Bohr radius of the muonic hydrogen is about 200 times smaller. As a result, the probability to find the muon in the nucleus is a factor $(m_\mu/m_e)^3 \sim 10^7$ larger than for the electron. The muonic hydrogen, therefore, has a much a larger sensitivity to nuclear structure, and to short-range effects in general.

10.1 Quantum-mechanical Coulomb problem

The starting point is, of course, the Coulomb problem solved by using either the Dirac or Schrödinger equation [1]. To set the stage, we recall that the hydrogen energy spectrum

$$E_n = -(Z\alpha)^2 m_r/2n^2, \tag{10.1}$$

with $m_r = mM/(M+m)$, where m, M are the masses of the lepton and the nucleus, as well as the corresponding wave functions $\phi_{nlm}(\boldsymbol{r})$ are obtained from the Schrödinger equation with the Coulomb potential:

$$\left[-\frac{\nabla^2}{2m_r} + V_C(r) - E_n\right]\phi_{nlm}(\boldsymbol{r}) = 0, \qquad V_C(r) = -\frac{Z\alpha}{r}, \tag{10.2}$$

doi:10.1088/978-0-7503-3431-0ch10

or, in momentum space, from the homogeneous Lippmann–Schwinger equation:

$$\left(\frac{\boldsymbol{p}^2}{2m_\mathrm{r}} - E_\mathrm{n}\right)\varphi_\mathrm{nlm}(\boldsymbol{p}) = \int \frac{d\boldsymbol{p}'}{(2\pi)^3} V_\mathrm{C}(|\boldsymbol{p} - \boldsymbol{p}'|)\, \varphi_{nlm}(\boldsymbol{p}'), \qquad V_\mathrm{C}(|\boldsymbol{q}|) = -\frac{4\pi Z\alpha}{\boldsymbol{q}^2}. \quad (10.3)$$

The coordinate and momentum representations are related via the three-dimensional Fourier transform. To compute perturbative effects due to a small correction $V_\varepsilon(\boldsymbol{p}' - \boldsymbol{p}; \boldsymbol{p}', \boldsymbol{p})$, the following matrix elements are required:

$$\langle n'l'm'|\, V_\varepsilon\, |nlm\rangle = \int \frac{d\boldsymbol{p}\, d\boldsymbol{p}'}{(2\pi)^3} V_\varepsilon(\boldsymbol{p}' - \boldsymbol{p}; \boldsymbol{p}', \boldsymbol{p})\, \varphi^*_{n'l'm'}(\boldsymbol{p}')\, \varphi_{nlm}(\boldsymbol{p}). \quad (10.4)$$

For a central potential, $V_\varepsilon = V_\varepsilon(|\boldsymbol{p}' - \boldsymbol{p}|)$,

$$\langle n'l'm'|\, V_\varepsilon\, |nlm\rangle = \delta_{ll'}\delta_{mm'} \int_0^\infty \frac{d|\boldsymbol{q}|\, \boldsymbol{q}^2}{2\pi^2} V_\varepsilon(|\boldsymbol{q}|)\, w_{nl}(\boldsymbol{q}^2), \quad (10.5)$$

$$w_{nl}(\boldsymbol{q}^2)\, Y^*_{lm}(\Omega_\mathrm{q}) \equiv \int d\boldsymbol{p} \varphi^*_{nlm}(\boldsymbol{p} + \boldsymbol{q})\, \varphi_{nlm}(\boldsymbol{p}). \quad (10.6)$$

For a zero-range correction, V_ε is constant and we are left with

$$\frac{1}{(2\pi)^3} \int d\boldsymbol{p}\ \varphi_{nlm}(\boldsymbol{p}) \int d\boldsymbol{p}'\ \varphi^*_{n'l'm'}(\boldsymbol{p}') = \phi_{nlm}(0)\, \phi_{n'l'm'}(0) = \frac{(Z\alpha m_\mathrm{r})^3}{\pi(nn')^{3/2}} \delta_{l'0}\delta_{m'0}\delta_{l0}\delta_{m0}. \quad (10.7)$$

10.2 One-photon exchange in dispersive representation

While the finite-size contribution is increased by the aforementioned factor of 10^7, relative to normal atoms, the QED effects contributing to the $2S$–$2P$ splitting increase only by a factor of 50, promoting the finite-size contribution to be the second largest contribution, trumped only by the one-loop electron vacuum-polarization (eVP), shown in figure 10.1(a). The literature accounting for these effects is very extensive, see e.g. [2–6]. Nevertheless, the work on accurate calculations of these effects will continue in the foreseeable future, as the ongoing experiments bode new leaps in precision. Important for this program is the progress on the nuclear side, since many of the corrections require the input of nuclear and nucleon form factors and structure functions.

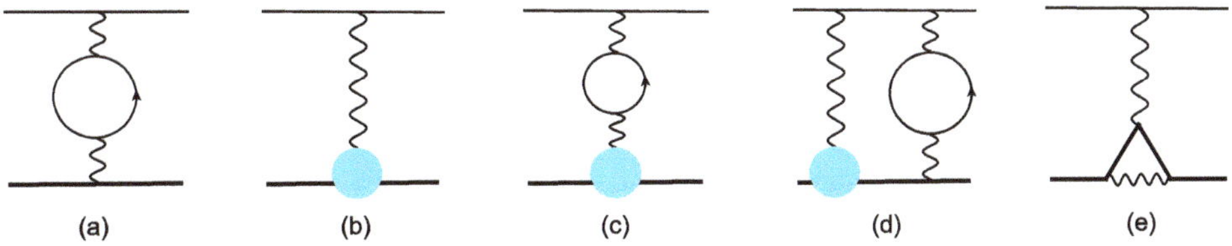

Figure 10.1. Major corrections discussed in the text. The cyan blobs represent the finite-size effects, and the thin and thick lines represent the muon and proton, respectively.

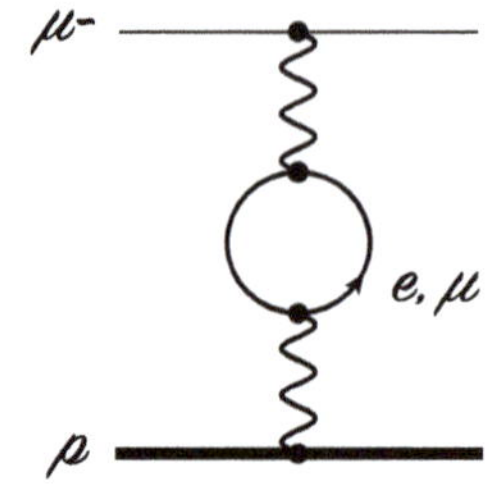

Figure 10.2. One-loop electron or muon vacuum-polarization in the one-photon-exchange potential of muonic hydrogen.

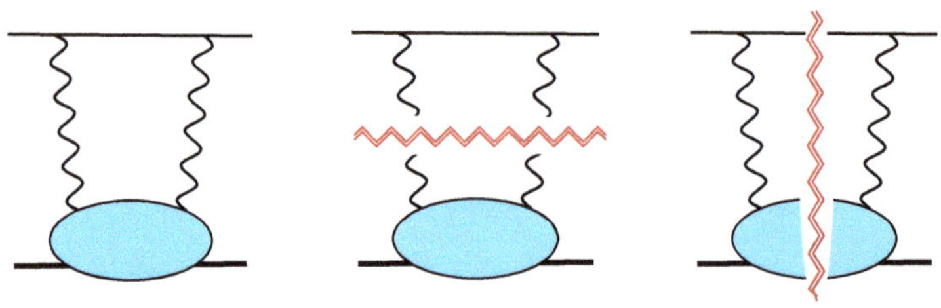

Figure 10.3. (a) The two-photon exchange, with (b) the *t*-channel and (c) the *s*-channel cuts. The cyan blobs represent effects from nuclear excitations.

The Coulomb-potential problem, described above, assumes a point-like nucleus with the electric charge Ze, whereas the effects beyond this approximation come as perturbative corrections to the Lamb shift, fine and hyperfine structure. The perturbation series is organized in powers of the fine-structure constant $\alpha = e^2/4\pi$, and mass ratios. Among the nuclear-structure effects, one distinguished (i) the *finite-size* effects, which come from the electromagnetic distributions in the nucleus (e.g. figure 10.1(b–d)) and (ii) the *polarizability* effects (figure 10.2(a)), caused by deformations of the distributions within the atom. The former can be entirely described by the elastic form factors, such as $G_E(Q^2)$ and $G_M(Q^2)$ in the case of a spin-1/2 nucleus (e.g. the proton), whereas the latter requires a more complicated input, *viz*, structure functions. The two types of effects are also treated quite differently, as will be seen in the following derivation of the leading and next-to-leading nuclear contributions. Here we shall see how some of these ingredients are entering the atomic calculations (figure 10.3).

10.3 Vacuum polarization contributions to the Lamb shift

10.3.1 The first-order effect

Consider first the effect of vacuum-polarization in first-order time-independent perturbation theory. As established in chapter 9, the photon propagator modified by vacuum-polarization is given by

$$-\frac{ig^{\mu\nu}}{q^2[1-\Pi(q^2)]} = -\frac{ig^{\mu\nu}}{q^2}[1+\Pi(q^2)+\Pi^2(q^2)+\cdots], \tag{10.8}$$

where the first term on the right-hand side contains the Coulomb potential, and the others are corrections to it. Using the once-subtracted dispersion relation,

$$\Pi(q^2) = \frac{1}{\pi}\int_{s_0}^{\infty} ds\, \frac{\mathrm{Im}\,\Pi(s)}{s-q^2-i0^+}\frac{q^2}{s}, \tag{10.9}$$

the first correction to the Coulomb potential is described by the Uehling potential [7, 8]:

$$V_{\mathrm{U}}(r) = \frac{Z\alpha}{\pi r}\int_{s_0}^{\infty} ds\, e^{-r\sqrt{s}}\,\frac{\mathrm{Im}\,\Pi(s)}{s}. \tag{10.10}$$

Its first-order correction to the $2P-2S$ Lamb shift is easily computed as

$$\begin{aligned} E_{\mathrm{U}}(2P-2S) &= \int d\vec{r}\,\left(|\phi_{210}(\vec{r})|^2 - |\phi_{200}(\vec{r})|^2\right)V_{\mathrm{U}}(r) \\ &= -\frac{(Z\alpha)^4 m_{\mathrm{r}}^3}{\pi}\int_{s_0}^{\infty} ds\,\frac{\mathrm{Im}\,\Pi(s)}{(\sqrt{s}+\alpha m_{\mathrm{r}})^4}. \end{aligned} \tag{10.11}$$

Substituting the expression for the one-loop eVP in QED, which we encountered in equation (9.4), we find the following expression for the muonic hydrogen:

$$\begin{aligned} \Delta E^{(1)1\mathrm{loop}}(2P-2S) = \frac{\alpha^3 m_{\mathrm{r}}}{\kappa^3}\Bigg(&\frac{1}{3} + \frac{\kappa(2\kappa^6 - 13\kappa^4 + 44\kappa^2 - 24)}{36\pi(1-\kappa^2)^2} \\ &- \frac{15\kappa^4 - 20\kappa^2 + 8}{12\pi(1-\kappa^2)^{5/2}}\arccos\kappa\Bigg) \simeq 205.0074\ \mathrm{meV}, \end{aligned} \tag{10.12}$$

where $\kappa = Z\alpha m_{\mathrm{r}}/2m_{\mathrm{e}} \simeq 0.6781$. In this example, the integrand cannot be expanded in $Z\alpha$, since κ is of order 1, i.e. the Bohr radius is comparable with the Compton wavelength of the virtual e^+e^- pair. For this reason, the Uehling correction is effectively of order $\alpha(Z\alpha)^2$, rather than $\alpha(Z\alpha)^4$ as suggested by naive counting.

The two-loop eVP is somewhat more complicated:

$$\begin{aligned} \Pi^{2\mathrm{loop}}(s) = -\frac{\alpha^2}{3\pi}\Bigg\{ &\frac{1}{2}(3-v^2)(1+v^2)\Bigg[\frac{\pi^2}{6} + \log\frac{1+v}{1-v}\log\frac{1+v}{2} + 2\,\mathrm{Li}_2\!\left(\frac{1-v}{1+v}\right) \\ &+ 2\,\mathrm{Li}_2\!\left(\frac{1+v}{2}\right) - 2\,\mathrm{Li}_2\!\left(\frac{1-v}{2}\right) - 4\,\mathrm{Li}_2(v) + \mathrm{Li}_2(v^2)\Bigg] \\ &+ \frac{3}{8}v(5-3v^2) - v(3-v^2)[\log 8v^2 - 3\log(1+v)] \\ &+ \frac{1}{16}(33 - 72v + 22v^2 + 24v^3 - 7v^4)\log\frac{1+v}{1-v}\Bigg\}, \end{aligned} \tag{10.13}$$

where $v = \sqrt{1 - \frac{4m^2}{s}}$. The corresponding contribution to the Lamb shift is referred to as Kallen–Sabry contribution. Its numerical value for muonic hydrogen is 1.253 00 meV.

To account for the iteration of the one-loop eVP, i.e. the 1PI-reducible two-loop effect (see the third term on the right-hand side of equation (10.8)), we need as well the real part of the eVP at one loop:

$$\Pi^{1\text{loop}}(t) = \frac{\alpha}{3\pi}\left\{\frac{1}{3} + 2\left(1 + \frac{2m^2}{t}\right)\left[\sqrt{1 - \frac{4m^2}{t}}\ \text{arccoth}\left(\sqrt{1 - \frac{4m^2}{t}}\right) - 1\right]\right\}. \quad (10.14)$$

The one-loop iteration is found as

$$2\,\text{Re}\,\Pi^{1\text{loop}}(t) \times \text{Im}\,\Pi^{1\text{loop}}(t), \quad (10.15)$$

which amounts to 0.255 meV in the Lamb shift of μH. Adding this effect, to the irreducible two-loop contribution, one obtains

$$\Delta E^{(1)2\text{loop}}(2P - 2S) = 1.507\,95\ \text{meV}. \quad (10.16)$$

10.3.2 Second-order effect

At the second-order perturbation theory the energy shift is given by

$$\delta E_{2l}^{(2)} = \sum_{n\neq 2} \frac{|\langle 2lm|\ V_{\text{U}}\ |nlm\rangle|^2}{E_2^{(0)} - E_{\text{n}}^{(0)}}, \quad (10.17)$$

where the integral over the continuum spectrum is understood.

The discrete spectrum, for the one-loop eVP in μH gives

$$\delta E_{2S}^{(2)1\text{loop}} = 0.217\,19 - 0.040\,18 - 0.012\,50 - 0.005\,70 - \dots = 0.148\,32\ \text{meV}, \quad (10.18)$$

$$\delta E_{2P}^{(2)1\text{loop}} = -0.000\,35\ \text{meV}. \quad (10.19)$$

The matrix elements for the continuum spectrum can be obtained from the discrete spectrum with the use of the following replacement,

$$n \to 1/ik, \quad (10.20)$$

and the following identities

$$\frac{(1/2 - ik + u/\kappa)^{1/ik}}{(1/2 + ik + u/\kappa)^{1/ik}} = \exp\left[\frac{1}{ik}\log\frac{1/2 - ik + u/\kappa}{1/2 + ik + u/\kappa}\right] = \exp\left[-\frac{2}{k}\arctan\frac{k}{1/2 + u/\kappa}\right]. \quad (10.21)$$

As the result one obtains

$$\delta E_{2S}^{(2)\text{cont.}} = -\frac{4Z\alpha}{\pi^2 a_B}\int_0^\infty \frac{dk}{2\pi}\frac{2}{(1/4+k^2)}\frac{2\pi k\, e^{\pi/k}}{\sinh(\pi/k)}$$
$$\times\left|\int_1^\infty \frac{du}{u}\frac{(u/\kappa)^2+k^2+3/4}{[(1/2+u/\kappa)^2+k^2]^2}\exp\left[-\frac{2}{k}\arctan\frac{k}{1/2+u/\kappa}\right]\text{Im}\,\Pi(4m^2u^2)\right|^2 \quad (10.22)$$
$$\simeq -0.301\,48\text{ meV},$$

and

$$\delta E_{2P}^{(2)\text{cont.}} = -\frac{4Z\alpha}{\pi^2 a_B}\int_0^\infty dk\,\frac{2}{(1/4+k^2)}\frac{k(k^2+1)\,e^{\pi/k}}{3\sinh(\pi/k)}$$
$$\times\left|\int_1^\infty \frac{du}{u}\frac{1}{[(1/2+u/\kappa)^2+k^2]^2}\exp\left[-\frac{2}{k}\arctan\frac{k}{1/2+u/\kappa}\right]\text{Im}\,\Pi(4m^2u^2)\right|^2 \quad (10.23)$$
$$\simeq -0.001\,92\text{ meV}.$$

The total second-order one-loop eVP contribution to the Lamb shift of μH is given by

$$E^{(2)}(2P-2S) \simeq 0.150\,89\text{ meV}, \quad (10.24)$$

which coincides with the values given in the literature [2].

The generic feature of the second-order effect is that the levels with the same quantum numbers l and m repel. The ground state therefore pushes $2S$ up thus reducing the Lamb shift. The upper S-levels push $2S$ back down and their collective effect is about twice larger than the effect of the ground state. The second-order effect on the p-level is relatively small.

10.4 Finite-size effects

Let us start by discussing the finite-size effects, that can be described through the charge, Friar and Zemach radii, as well as higher moments of the electromagnetic distributions.

10.4.1 Lamb shift

The main nuclear effect in the Lamb shift comes from the nuclear charge distribution $\rho_E(r)$, which, in momentum space, is described by an electric form factor (eFF) $G_E(Q^2)$, function of the photon virtuality, $Q^2 = \boldsymbol{q}^2 - q_0^2$. The corresponding potential is (see figure 10.1(b))

$$V_{\text{eFF}}(|\boldsymbol{q}|) = -\frac{4\pi Z\alpha}{\boldsymbol{q}^2}[G_E(\boldsymbol{q}^2)-1] = \frac{1}{\pi}\int_{t_0}^\infty \frac{dt}{t}\,\text{Im}\,G_E(t)\,\frac{4\pi Z\alpha}{\boldsymbol{q}^2+t}, \quad (10.25)$$

neglecting the relativistic effects, such as the dependence on the energy transfer q_0 (retardation) and recoil corrections, which can be treated within the Breit-potential formalism, see [9, chapter IX, §83]. The two forms of the potential in equation (10.25) are related by the once-subtracted dispersion relation:

$$G_{\rm E}(Q^2) = 1 - \frac{Q^2}{\pi}\int_{t_0}^{\infty}\frac{dt}{t}\frac{\operatorname{Im} G_{\rm E}(t)}{t + Q^2 - i0^+}, \tag{10.26}$$

where the integration is done over the time-like region, where the form factor develops a discontinuity associated with particle production, with t_0 being the lowest threshold. For the proton, for instance, the leading discontinuity is associated with two-pion production, i.e. $t_0 = 4m_\pi^2$.

In principle, the absorptive part of the form factor, $\operatorname{Im} G_{\rm E}(t)$, is known empirically (see e.g. [10–13] for the proton). However, here we use the dispersive representation simply as a convenient analytical tool. In coordinate space, where the Coulomb potential is given by $-Z\alpha/r$, the form-factor correction takes the following form,

$$V_{\rm eFF}(r) = \frac{Z\alpha}{\pi}\int_{t_0}^{\infty}\frac{dt}{t}\operatorname{Im} G_{\rm E}(t)\,\frac{1}{r}e^{-r\sqrt{t}}, \tag{10.27}$$

which, in fact, is the Yukawa potential with the dispersed mass given by $\sqrt{t}$. As a result, the first-order perturbation-theory contribution to the classic $(2S - 2P)$ Lamb shift is given by

$$E_{2S-2P}^{\langle{\rm eFF}\rangle} \equiv \langle 2S|V_{\rm eFF}|2S\rangle - \langle 2P|V_{\rm eFF}|2P\rangle = \frac{(Z\alpha)^4 m_{\rm r}^3}{2\pi}\int_{t_0}^{\infty} dt\,\frac{\operatorname{Im} G_{\rm E}(t)}{(\sqrt{t} + Z\alpha m_{\rm r})^4}. \tag{10.28}$$

Since $Z\alpha m_{\rm r} \ll \sqrt{t_0}$, the denominator can be expanded (assuming the atomic Bohr radius is much larger than the nuclear size), yielding:

$$E_{2S-2P}^{\langle{\rm eFF}\rangle} = \frac{(Z\alpha)^4 m_{\rm r}^3}{12}\sum_{k=2}^{\infty}\frac{(-Z\alpha m_{\rm r})^{k-2}}{(k-2)!}\langle r_{\rm E}^k\rangle = \frac{(Z\alpha)^4 m_{\rm r}^3}{12}(\langle r_{\rm E}^2\rangle - Z\alpha m_{\rm r}\langle r_{\rm E}^3\rangle + \cdots), \tag{10.29}$$

where $\langle r_{\rm E}^k\rangle$ is the kth moment of the charge distribution $\rho_{\rm E}(r)$:

$$\langle r_{\rm E}^k\rangle = 4\pi\int_0^{\infty} dr\, r^{2+k}\rho_{\rm E}(r) = \frac{(k+1)!}{\pi}\int_{t_0}^{\infty} dt\frac{\operatorname{Im} G_{\rm E}(t)}{t^{1+k/2}}. \tag{10.30}$$

The first term in equation (10.29) is the famous charge-radius correction, appearing at order $(Z\alpha)^4$. To compute the next term of order $(Z\alpha)^5$ correctly, we ought to take this correction to the second-order perturbation theory:

$$\begin{aligned} E_{2S-2P}^{\langle{\rm eFF}\rangle\langle{\rm eFF}\rangle} &= \sum_{n\neq 2}\frac{|\langle 2S|V_{\rm eFF}|nS\rangle|^2 - |\langle 2P|V_{\rm eFF}|nP\rangle|^2}{E_2 - E_n} \\ &\cong -\frac{2(Z\alpha)^5 m_{\rm r}^4}{\pi}\int_0^{\infty}\frac{dQ}{Q^4}[G_{\rm E}(Q^2) - 1]^2 = \frac{(Z\alpha)^5 m_{\rm r}^4}{12}\left(\langle r_{\rm E}^3\rangle - \frac{1}{2}\langle r_{\rm E}^3\rangle_{(2)}\right), \end{aligned} \tag{10.31}$$

where we only kept terms of order $(Z\alpha)^5$ and introduced the third Zemach moment [14]:

$$\langle r_E^3\rangle_{(2)} = \frac{48}{\pi}\int_0^\infty \frac{dQ}{Q^4}\left[G_E^2(Q^2) - 1 + \frac{1}{3}\langle r_E^2\rangle Q^2\right], \tag{10.32}$$

with the corresponding radius called the Friar radius, $r_{\text{Friar}} = \sqrt[3]{\langle r_E^3\rangle_{(2)}}$.

One sees that the second-order contribution essentially replaces the third radius, appearing in equation (10.29), with the Friar-radius term. To order $(Z\alpha)^5$, the finite-size correction is then given by

$$E_{2S-2P}^{\text{f.s.}} = \frac{(Z\alpha)^4 m_r^3}{12}\langle r_E^2\rangle - \frac{(Z\alpha)^5 m_r^4}{24}\langle r_E^3\rangle_{(2)}. \tag{10.33}$$

Similarly, to compute the complete order-$(Z\alpha)^6$ correction, one needs to take this potential to the third-order, and so forth. The first corrections that affect P-levels begin at order $(Z\alpha)^6$. Thus, up to this order the entire effect can be deduced from a $\delta(\boldsymbol{r})$-function potential, which provides an easy generalization for the nS-level shift:

$$E_{nS}^{\text{f.s.}} = \frac{2\pi}{3}Z\alpha\left(\langle r_E^2\rangle - \frac{1}{2}Z\alpha m_r\langle r_E^3\rangle_{(2)}\right)\phi_{nS}^2(0), \tag{10.34}$$

with $\phi_{nS}^2(0)$ the wave function at the origin. The Friar-radius contribution is obviously playing a more significant role in μH than in H, and was an initial suspect to resolve the proton-radius puzzle [15]. However, in that scenario, the Friar radius was so large that the expansion in radii would be invalidated [16, 17]. The present consensus is that this contribution is at least an order of magnitude smaller than what is required for the explanation of the puzzle. Furthermore, there are relativistic corrections, which can be treated within the Breit-potential formalism or, alternatively, by considering the two-photon exchange, as will be seen below. Also important are some radiative corrections, which come from combining the finite-size and QED effects. In muonic atoms, the eVP plays an especially prominent role, and produces sizeable radiative corrections to the finite-size effects shown in figures 10.1(c) and (d), considered in section 10.6.

10.4.2 Hyperfine splitting

Assuming a spin-1/2 nucleus, the hyperfine splitting (HFS) arises from the magnetization properties of the nucleus described by the magnetic form factor (mFF) $G_M(Q^2)$. For the S-levels, the corresponding potential is given by (omitting recoil corrections):

$$\begin{aligned} V_{\text{mFF}}^F(|\boldsymbol{q}|) &= \frac{4\pi Z\alpha}{3mM}\left[F(F+1) - \frac{3}{2}\right]G_M(\boldsymbol{q}^2) \\ &= \frac{4Z\alpha}{3mM}\left[F(F+1) - \frac{3}{2}\right]\int_{t_0}^\infty dt\, \frac{\text{Im}\, G_M(t)}{\boldsymbol{q}^2 + t}, \end{aligned} \tag{10.35}$$

where $F = 0$ or 1 is the total spin, κ_N the anomalous magnetic moment of the nucleus, and $G_M(0) = 1 + \kappa_N$ is the value of the magnetic moment in units of $Ze/2M$. The corresponding coordinate-space potential is directly proportional to the magnetization density $\rho_M(r)$.

The first-order contribution yields the following HFS interval of the nS-level:

$$E_{nS-\text{hfs}}^{\langle\text{mFF}\rangle} = (1 - 2Z\alpha m_r \langle r_M \rangle)\frac{E_F}{n^3} + O[(Z\alpha)^6], \tag{10.36}$$

with $E_F = \frac{8(Z\alpha)^4 m_r^3 (1+\kappa_N)}{3mM}$ the Fermi energy, and $\langle r_M \rangle = 4\pi \int_0^\infty dr\, r^3 \rho_M(r)$ the linear magnetic radius. At the second order, the interference with the eFF potential of equation (10.25) gives

$$E_{nS-\text{hfs}}^{\langle\text{mFF}\rangle\langle\text{eFF}\rangle} = Z\alpha m_r (\langle r_M \rangle - r_Z)\frac{E_F}{n^3} + O[(Z\alpha)^6], \tag{10.37}$$

thus canceling the linear magnetic radius term from the first order, and installing instead the Zemach radius:

$$r_Z = -\frac{4}{\pi}\int_0^\infty \frac{dQ}{Q^2}\left[\frac{G_E(Q^2)G_M(Q^2)}{1+\kappa_N} - 1\right]. \tag{10.38}$$

The Fermi-energy contribution is not a finite-size effect, as it is already present for a point-like nucleus. The leading finite-size effect in the HFS is therefore of order $(Z\alpha)^5$,

$$E_{nS-\text{hfs}}^{\text{f.s.}} = -2Z\alpha m_r r_Z \frac{E_F}{n^3}. \tag{10.39}$$

At this order, also the polarizability corrections begin to appear. We consider them next.

10.5 Two-photon exchange and polarizability effects

Thus far, we considered effects which stem from the one-photon exchange and its iterations, such that the nucleus stays intact and in its ground state. There are also effects coming from nuclear excitations, which can only be assessed through a two-photon exchange, see figure 10.2(a). This description goes beyond the elastic form factors and involves instead the polarizabilities and inelastic structure functions, as will be seen in what follows.

The two-photon exchange in figure 10.2(a) introduces, in general, a correction $V_{2\gamma}(p' - p; p', p)$ which depends on the relative momenta of the initial and final state, p and p', as well as the momentum transfer $q = p' - p$. These are four-momenta, but the energy effects can safely be neglected, since they are suppressed by $(Z\alpha)^2 m_r$. The dependence on $|\boldsymbol{p}| = |\boldsymbol{p}'|$ is suppressed by $Z\alpha m_r$ and will, to leading order, be neglected as well. The dependence on $|\boldsymbol{q}|$ is a bit more subtle, because of possible $\log Z\alpha$ terms. To see this, let us write a dispersion relation for the graph of figure 10.2(b) in the Mandelstam variable $t = q^2 = -Q^2$, for fixed $s \simeq (m + M)^2$:

$$V_{2\gamma}(t) = \frac{1}{\pi}\int_0^\infty dt' \, \frac{\text{Im } V_{2\gamma}(t')}{t' - t - i0^+} = V_{2\gamma}(t=0) + \frac{t}{\pi}\int_0^\infty dt' \, \frac{\text{Im } V_{2\gamma}(t')}{t'(t' - t - i0^+)}, \quad (10.40)$$

where in the second equation we have a once-subtracted relation, thus introducing the forward scattering amplitude, $V_{2\gamma}(t = 0)$. The remainder with non-vanishing momentum transfer is referred to as the off-forward amplitude.

The absorptive part, Im $V_{2\gamma}(t)$, corresponds to the discontinuity across the t-channel cut, which starts at 0, because the photons are massless. Because of this, the expansion in t is non-analytic. When calculating the level shifts in perturbation theory, this non-analyticity translates into $\log Z\alpha$ contributions.

Let us see this for the Lamb shift, where we can make use of the arguments leading to equation (10.28) and arrive at

$$E_{2S-2P}^{\langle 2\gamma\rangle} = \frac{(Z\alpha m_\mathrm{r})^3}{8\pi^2}\int_0^\infty dt \, \frac{t \text{ Im } V_{2\gamma}(t)}{(\sqrt{t} + Z\alpha m_\mathrm{r})^4} \quad (10.41\text{a})$$

$$= V_{2\gamma}(0)\, \phi_{2S}^2(0) + \frac{(Z\alpha m_\mathrm{r})^3}{8\pi^2}\int_0^\infty dt \left[\frac{t}{(\sqrt{t} + Z\alpha m_\mathrm{r})^4} - \frac{1}{t}\right] \text{Im } V_{2\gamma}(t). \quad (10.41\text{b})$$

Let us focus on the leading polarizability effect, coming from the electric α_{E1} and magnetic β_{M1} dipole polarizabilities of the nucleus. Knowing how they enter the Compton scattering amplitude, we can obtain their contribution to Im $V_{2\gamma}(t)$. Note that our consideration of the two-photon cut in figure 10.2(b) involves only the real Compton scattering and static polarizabilities. The expression is rather lengthy [18] and we only quote here the polarizability potential in the well-known long-range and the singular short-range limit:

$$V_{2\gamma}(r) = \frac{1}{4\pi^2 r}\int_0^\infty dt \text{ Im } V_{2\gamma}(t)\, e^{-r\sqrt{t}} \overset{r\to\infty}{=} -\frac{\alpha\alpha_{\mathrm{E1}}}{2r^4} + \frac{\alpha(11\alpha_{\mathrm{E1}} + 5\beta_{\mathrm{M1}})}{4\pi m r^5} + O(1/r^6), \quad (10.42\text{a})$$

$$\overset{r\to 0}{=} \frac{\alpha m \log(mr)}{2\pi r^3}(\alpha_{\mathrm{E1}} - \beta_{\mathrm{M1}}) + O(1/r^3). \quad (10.42\text{b})$$

Note that the prefactor of α is coming from the lepton line, whereas the polarizabilities contain $Z^2\alpha$, hence in total the order is $(Z\alpha)^2$ as expected. The entire potential is negative-definite (attractive), provided $\alpha_{\mathrm{E1}} > \beta_{\mathrm{M1}}$, and assuming the electric polarizability is a positive-definite quantity. This potential, however, is not very useful to compute the contribution to the S-states, because of the singular short-range behavior. Anticipating that, we have introduced the subtracted dispersion relation in equation (10.40), leading to equation (10.41b). The integration over t is now convergent, the short-range contribution regularized, and, for the leading-$Z\alpha$ contribution to the nS-shift, we obtain

$$E_{nS}^{\langle 2\gamma\rangle} = \left\{V_{2\gamma}(0) + 4\alpha(Z\alpha)\,\alpha_{\mathrm{E1}} \log\left[2n(Z\alpha)^{-1}\right] + O[(Z\alpha)^3]\right\} \phi_{nS}^2(0). \quad (10.43)$$

It remains now to calculate $V_{2\gamma}(0)$, i.e. the forward two-photon exchange, as it apparently gives the larger contribution, of order $(Z\alpha)^5$. For this, one can make use of the s-channel dispersion relations for the forward Compton amplitude, see figure 10.2(c), which allows one to express everything in terms of integrals of the structure functions, $F_{1,2}(x, Q^2)$, measured in inclusive electron scattering. Unfortunately, one of these dispersion relations requires a subtraction too, which precludes the use of a purely data-driven approach. Still, most of the existing calculations are based on the data-driven approach, in conjunction with some model-building of the subtraction function. Alternatively, one can calculate the entire $(Z\alpha)^5$ contribution using χPT or lattice QCD.

Similar consideration can be done for the HFS. There are two important differences: the contribution of order $(Z\alpha)^6 \log Z\alpha$ is absent, and the order $(Z\alpha)^5$—the forward two-photon-exchange contribution—can be expressed in terms of the spin structure functions without a subtraction. There is an interesting order-$(Z\alpha)^6$ contribution coming from the neutral-pion exchange, which couples to the lepton through the chiral anomaly [19–22]. However, it is not very relevant at the current level of precision.

10.6 Radiative corrections

The largest corrections in muonic atoms involve the eVP, which also produces radiative corrections to the finite-size effects via the mechanisms of figures 10.1(c) and (d). They are, respectively, referred to as VP1 and VP2.

10.6.1 VP2 correction

The diagram in figure 10.1(d) appears from the interference of the finite-size correction V_{eFF}, figure 10.1(b), with the Uehling potential, figure 10.1(a), at the second-order perturbation theory. To avoid lengthy considerations, we cast the finite-size effects into a $\delta(\boldsymbol{r})$-function potential, as remarked above (valid for contributions that only influence the S-levels). The eVP effect then amounts to correcting the wave function at the origin, appearing in equation (10.34), as follows:

$$\phi_{nS}^2(0) \to \phi_{nS}^2(0)\left[1 + \frac{\alpha}{\pi}\, C_1(nS)\right], \tag{10.44}$$

where $C_1(nS)$ is known analytically for the case of one-loop eVP, see [23, equation (B3)]. This is the universal eVP correction to any δ-function potential, including the forward two-photon-exchange correction.

10.6.2 VP1 correction to the Lamb shift

This correction (figure 10.1(c)) corresponds to the following potential,

$$V_{\text{eFF,VP1}}(|\boldsymbol{q}|) = -\frac{4\pi Z\alpha}{\boldsymbol{q}^2}[G_E(\boldsymbol{q}^2) - 1]\Pi(\boldsymbol{q}^2), \tag{10.45}$$

where $\Pi(Q^2)$ is the scalar part of the vacuum-polarization. Using the dispersive representation, the correction to equation (10.28) reads

$$\begin{aligned} E^{\mathrm{VP1}}_{2S-2P} &= \frac{(Z\alpha)^4 m_\mathrm{r}^3}{2\pi}\left[\int_{t_0}^{\infty} dt\, \frac{\mathrm{Im}\, G_\mathrm{E}(t)\, \mathrm{Re}\,\Pi(-t)}{(\sqrt{t}+Z\alpha m_\mathrm{r})^4} + \int_{4m_e^2}^{\infty} dt\, \frac{[\mathrm{Re}\, G_\mathrm{E}(-t)-1]\, \mathrm{Im}\,\Pi(t)}{(\sqrt{t}+Z\alpha m_\mathrm{r})^4}\right] \\ &= \frac{(Z\alpha)^4 m_\mathrm{r}^3}{2\pi}\left[\int_{4m_e^2}^{\infty} dt\, \underbrace{[\mathrm{Re}\, G_\mathrm{E}(-t)-1]}_{\approx \frac{1}{6}t\langle r_\mathrm{E}^2\rangle}\, \mathrm{Im}\,\Pi(t)\left(\frac{1}{(\sqrt{t}+Z\alpha m_\mathrm{r})^4} - \frac{1}{t^2}\right) + O\left(\frac{Z\alpha m_\mathrm{r}}{\sqrt{t_0}}\right)\right]. \end{aligned} \tag{10.46}$$

Within the indicated approximations this correction affects only the charge-radius contribution:

$$\begin{aligned} E^{\mathrm{f.s.}}_{2S-2P} &= \frac{1}{12}(Z\alpha)^4 m_\mathrm{r}^3\left[(1+\delta^{\mathrm{VP1}}_{2S-2P})\langle r_\mathrm{E}^2\rangle - \frac{1}{2}Z\alpha m_\mathrm{r}\langle r_\mathrm{E}^3\rangle_{(2)}\right] \\ \delta^{\mathrm{VP1}}_{2S-2P} &= \frac{1}{\pi}\int_{4m_e^2}^{\infty} dt\, t\left[\frac{1}{(\sqrt{t}+Z\alpha m_\mathrm{r})^4} - \frac{1}{t^2}\right]\mathrm{Im}\,\Pi(t). \end{aligned} \tag{10.47}$$

Substituting the one-loop expression for the eVP from equation (9.4), we obtain

$$\delta^{\mathrm{VP1(1)}}_{2S-2P} = \frac{\alpha\kappa}{6\pi(1-\kappa^2)^2}\left[\kappa(4\kappa^2-7) + \frac{4\kappa^4-10\kappa^2+9}{\sqrt{1-\kappa^2}}\arccos\kappa\right], \quad \text{with } \kappa = \frac{Z\alpha m_\mathrm{r}}{2m_\mathrm{e}}. \tag{10.48}$$

For μH, $\delta^{\mathrm{VP1(1)}}_{2S-2P} \simeq 2.155 \times 10^{-3}$. Note that this correction affects both the S- and P-levels.

10.6.3 VP1 correction to HFS (figure 10.1(c))

The eVP radiative corrections to the HFS are treated similarly. In this case the potential corresponding to the diagram of figure 10.1(c) is

$$V^{\mathrm{hfs}}_{\mathrm{mFF,\,VP1}}(|\boldsymbol{q}|) = \frac{8\pi Z\alpha}{3mM}\, G_\mathrm{M}(\boldsymbol{q}^2)\,\Pi(\boldsymbol{q}^2). \tag{10.49}$$

Going through the same steps as in equation (10.46), one finds the following effect on the ground-state HFS:

$$\begin{aligned} E^{\mathrm{VP1}}_{1S-\mathrm{hfs}} = \frac{8(Z\alpha)^4 m_\mathrm{r}^3}{3\pi mM}&\left[\int_{4m_e^2}^{\infty} dt\, \underbrace{\mathrm{Re}\, G_\mathrm{M}(-t)}_{\approx(1+\kappa_N)}\, \mathrm{Im}\,\Pi(t)\left(\frac{1}{(\sqrt{t}+2Z\alpha m_\mathrm{r})^2} - \frac{1}{t}\right)\right. \\ &\left. - 4Z\alpha m_\mathrm{r}\int_{t_0}^{\infty} dt\, \frac{\mathrm{Im}\, G_\mathrm{M}(t)}{t^{3/2}}\, \mathrm{Re}\,\Pi(t) + O\left(\frac{(Z\alpha m_\mathrm{r})^2}{t_0}\right)\right]. \end{aligned} \tag{10.50}$$

The first term is not a finite-size correction as it only affects the Fermi-energy term, albeit differently for each nl. For the ground state, its effect is $E_\mathrm{F}(1+\delta^{\mathrm{VP1}}_{1S-\mathrm{hfs}})$, where the one-loop eVP gives

$$\begin{aligned}\delta_{1S-\text{hfs}}^{\text{VP1}} &= \frac{1}{\pi}\int_{4m_e^2}^{\infty} dt \left[\frac{1}{(\sqrt{t}+2Z\alpha m_r)^2} - \frac{1}{t}\right] \operatorname{Im}\Pi^{(1)}(t) \\ &= \frac{\alpha}{3\pi\kappa_1^3}\left[2\kappa_1 + \frac{1}{3}\kappa_1^3 + \frac{2-\kappa_1^2+2\kappa_1^4}{\sqrt{\kappa_1^2-1}}\operatorname{arccosh}\kappa_1 - \pi\right], \quad \text{with } \kappa_1 = \frac{Z\alpha m_r}{m_e}.\end{aligned} \tag{10.51}$$

For μH this amounts to about 2% correction to the Fermi energy or, in absolute terms, 0.374 65 meV. This is a fairly large effect, and one must consider the next term in equation (10.50), which eventually leads to a Zemach radius correction, $r_Z(1+\delta_{Z,1S}^{\text{VP1}})$.

10.6.4 Combining VP1 and VP2

We note that this formalism applies to all the vacuum-polarization contributions, including HVP, μVP in H, etc. However, in these cases one can expand in $Z\alpha$ before the t-integration, which simplifies things a lot. For example, the VP2 and VP1 corrections become equal at leading order, with their combined effect given by

$$\delta_{1S-\text{hfs}}^{\text{VP1+VP2}} = -\frac{8Z\alpha m_r}{\pi}\int_{t_0}^{\infty} dt\, \frac{\operatorname{Im}\Pi(t)}{t^{3/2}} + O(Z^2\alpha^3). \tag{10.52}$$

Plugging in the one-loop VP with a lepton with mass m_ℓ, or a charged scalar with mass m_π, one obtains $\frac{3}{4}Z\alpha^2(m_r/m_\ell)$ and $\frac{1}{8}Z\alpha^2(m_r/m_\pi)$, respectively. The latter result can be used to estimate the HVP contribution.

10.7 Proton self-energy and the charge-radius definition

Let us briefly consider the correction resulting from the difference between the self-energy of the bound and free proton, see figure 10.1(e). This contribution has the same topology as the finite-size effect, figure 10.1(b), hence, could in principle be absorbed into the proton form factors. The problem is that it differs slightly between H and μH and thus requires extra care. This difference is exactly the same for the electric and magnetic form factors (affects only the Dirac form factor), and is proportional to the logarithm of the reduced-mass ratio for the two hydrogens, i.e.

$$\operatorname{Im} G_E^{\text{H}-\mu\text{H}}(t) = \operatorname{Im} G_M^{\text{H}-\mu\text{H}}(t) = Z^2\alpha\frac{t-2M^2}{\sqrt{t(t-4M^2)}}\log\frac{m_r^{(\mu\text{H})}}{m_r^{(\text{H})}}. \tag{10.53}$$

Using this expression, one can match all of the aforementioned finite-size contributions. For instance, for the charge radius this correction would be

$$\begin{aligned} r_p^2(\text{H}) - r_p^2(\mu\text{H}) &= \frac{6}{\pi}\int_{4M^2}^{\infty} dt\, \frac{\operatorname{Im} G_E^{\text{H}-\mu\text{H}}(t)}{t^2} \\ &= \frac{2Z^2\alpha}{\pi M^2}\log\frac{m_r^{(\mu\text{H})}}{m_r^{(\text{H})}} = 0.0010737\ \text{fm}^2,\end{aligned} \tag{10.54}$$

which is at the level of the μH precision. Currently, this difference is already accounted for (in the charge radius only) by shifting these logarithms, accompanied by the Bethe logarithms on the proton line, into the QED contribution, see [2, 6]. This practice essentially establishes a unique definition of the charge radius of a bound proton.

The correction shown in figure 10.1(e) is infrared divergent for the free proton and hence the relation to the charge radius extracted from *ep* scattering depends on the treatment of radiative corrections in these experiments. As these radiative corrections on the proton line are known to be negligibly small [24], at least at the present level of precision of scattering experiments, a possible mismatch with the atomic definition can be ignored.

References

[1] Bethe H A and Salpeter E E 1957 *Quantum Mechanics of One- and Two-Electron Atoms* (Berlin: Springer)

[2] Pachucki K 1996 Theory of the Lamb shift in muonic hydrogen *Phys. Rev.* A **53** 2092–100

[3] Eides M I, Grotch H and Shelyuto V A 2001 Theory of light hydrogen-like atoms *Phys. Rep.* **342** 63–261

[4] Borie E 2012 Lamb shift in light muonic atoms: revisited *Ann. Phys.* **327** 733–63

[5] Indelicato P 2013 Nonperturbative evaluation of some QED contributions to the muonic hydrogen $n = 2$ Lamb shift and hyperfine structure *Phys. Rev.* A **87** 022501

[6] Karshenboim S G, Korzinin E Y, Shelyuto V A and Ivanov V G 2015 Theory of Lamb shift in muonic hydrogen *J. Phys. Chem. Ref. Data* **44** 031202

[7] Uehling E A 1935 Polarization effects in the positron theory *Phys. Rev.* **48** 55–63

[8] Schwinger J 1949 Quantum electrodynamics. II. Vacuum polarization and self-energy *Phys. Rev.* **75** 651–79

[9] Berestetskii V B, Lifshitz E M and Pitaevskii L P 1982 *Quantum Electrodynamics* (Course of Theoretical Physics vol 4) (Oxford: Butterworth-Heinemann)

[10] Lin Y-H, Hammer H-W and Meißner U-G 2022 New insights into the nucleon's electromagnetic structure *Phys. Rev. Lett.* **128** 052002

[11] Alarcón J M, Higinbotham D W, Weiss C and Ye Z 2019 Proton charge radius extraction from electron scattering data using dispersively improved chiral effective field theory *Phys. Rev.* C **99** 044303

[12] Lorenz I, Meißner U G, Hammer H-W and Dong Y-B 2015 Theoretical constraints and systematic effects in the determination of the proton form factors *Phys. Rev.* D **91** 014023

[13] Belushkin M A, Hammer H W and Meissner U G 2007 Dispersion analysis of the nucleon form-factors including meson continua *Phys. Rev.* C **75** 035202

[14] Friar J L 1979 Nuclear finite size effects in light muonic atoms *Annals Phys.* **122** 151

[15] De Rujula A 2010 QED is not endangered by the proton's size *Phys. Lett.* B **693** 555–8

[16] Distler M O, Bernauer J C and Walcher T 2011 The RMS charge radius of the proton and Zemach moments *Phys. Lett.* B **696** 343–7

[17] Hagelstein F and Pascalutsa V 2015 Breakdown of the expansion of finite-size corrections to the hydrogen Lamb shift in moments of charge distribution *Phys. Rev.* A **91** 040502(R)

[18] Hagelstein F 2017 Exciting nucleon in Compton scattering and hydrogen-like atoms *PhD thesis* Mainz University arXiv:1710.00874

[19] Hagelstein F and Pascalutsa V 2016 Proton structure in the hyperfine splitting of muonic hydrogen *PoS* **CD15** 077

[20] Huong N T, Kou E and Moussallam B 2016 Single pion contribution to the hyperfine splitting in muonic hydrogen *Phys. Rev.* D **93** 114005

[21] Zhou H-Q and Pang H-R 2015 One-pion-exchange effect in the energy spectrum of muonic hydrogen *Phys. Rev.* A **92** 032512
Zhou H-Q and Pang H-R 2016 Erratum: One-pion-exchange effect in the energy spectrum of muonic hydrogen *Phys. Rev.* A **93** 069903

[22] Dorokhov A E, Kochelev N I, Martynenko A P, Martynenko F A and Faustov R N 2017 The contribution of pseudoscalar mesons to hyperfine structure of muonic hydrogen *Phys. Part. Nucl. Lett.* **14** 857–64

[23] Karshenboim S G and Shelyuto V A 2021 Hadronic vacuum-polarization contribution to various QED observables *Eur. Phys. J.* D **75** 49

[24] Maximon L C and Tjon J A 2000 Radiative corrections to electron proton scattering *Phys. Rev.* C **62** 054320

www.ingramcontent.com/pod-product-compliance
Ingram Content Group UK Ltd.
Pitfield, Milton Keynes, MK11 3LW, UK
UKHW051849100926
8556IPUK00005B/535

9 780750 334297